Creating Environmental Business Value

Achieving Two Shades of Green

Creating Environmental Business Value

Achieving Two Shades of Green

Stephen Poltorzycki

Best wishes,

Crisp Publications

Editor-in-Chief: *William F. Christopher*

Project Editor: *Kathleen Barcos*

Editor: *Amy Marks*

Cover Design: *Kathleen Barcos*

Cover Production: *Russell Leong Design*

Book Design & Production: *London Road Design*

Printer: *Bawden Printing*

Library of Congress Card Catalog Number 98-73104

ISBN 1-56052-489-8

Contents

Introduction

The central concept of this book is that environmental value and business value represent "two shades of green." Business value involves those things important to a company's stakeholders. This shade of green represents the color of money–not as an end in and of itself, but as a way of measuring business value. Environmental value involves those things a company does to have less impact on the environment and to reduce risk. The shade of green represented here is that of environmental responsibility.

The key is that these are two shades of the *same color.* Both business and environmental managers can discover that there is a strong connection between business and environmental value. Business mangers will find that it is in their enlightened self interest to find the business value in environmental responsibility. Environmental managers will understand how to align environmental efforts in a way that promotes business value.

Ultimately, the two shades will blend together as business and environmental value become fully integrated. For most companies, this is a distant dream. The intent of this book is to help make this dream a reality.

That's the vision. But what's the current reality? Corporate environmental management has come a long way in the last two decades: evolving from an early focus on complying with regulatory requirements to a more recent risk-based focus as potential liabilities associated with acquisitions, divestitures, process changes, and day-to-day operations came under scrutiny. Moreover, in the last several years there has been a growing recognition of the potential for environmental contributions to improve operational efficiency and competitive advantage–in short, to deliver business value.

Realistically, most managers today would agree that environmental considerations are not yet fully integrated into business decisions. Many environmental managers are not yet steeped in the business implications of their initiatives, and may even blame business managers for not understanding the importance of environmental management. And business managers sometimes blame environmental managers for not communicating with them in a way that moves the business forward.

This book lays out the road map for realizing the business value of corporate environmental initiatives. The road map and its related insights grow out of work my Arthur D. Little colleagues and I have engaged in over the years to help companies integrate environmental value and business value.

Chapter 1 defines business value in ways both environmental managers and business managers can understand–by describing the links between environmental management, environmental economics, and business value. By understanding the full benefit and cost of environmental activities, business managers can appreciate how environmental management is tied to their company's self interest.

Chapters 2 through 6 lay out a five-step pathway for moving ahead, and also provide some important conceptual and operational tools. The basic steps along this five-step path include:

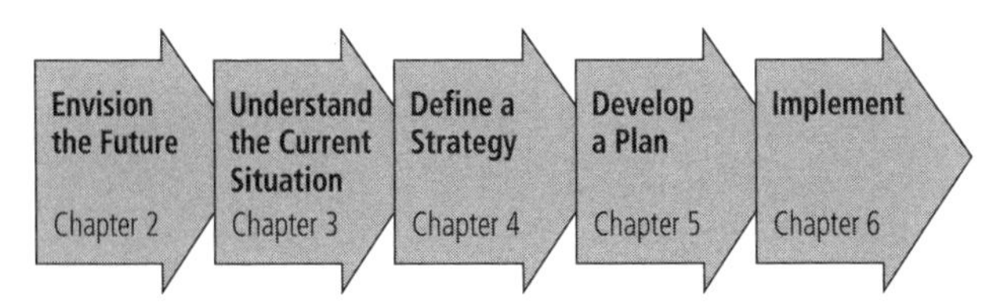

Chapter 2 describes the process of envisioning the future and answering the question, "What do we want to create?" Without a shared vision of how environmental management will align with business value, no progress will be possible.

Chapter 3 illustrates that, to realize a vision, it is essential to understand how the company's current situation stacks up to its vision. It provides tools for defining the nature of any gap that may exist between the vision and the current reality. The tension that arises by virtue of this gap creates the positive energy needed to move the company toward the vision.

Chapter 4 lays out the process for developing strategic options and selecting the right option to achieve the vision. Because there is more than one way to achieve a vision, the process of defining a strategy seeks to uncover the direction that is best for the company.

Chapter 5 shows how to develop a plan to operationalize the strategy. It uses the success some companies have had in creating a "Design for Environment" process as an example of successful integration of environmental thinking into business processes.

Chapter 6 makes the point that plans and strategies accomplish nothing by themselves. Once implemented, however, a new mission for environmental management will unfold that can lead to improved operational effectiveness and competitive advantage. Monsanto Company's success in implementing a bold new strategy based on a fresh understanding of how the environment can contribute to business success provides some helpful lessons.

The approach described in this book helps ensure that the pathway chosen for linking environmental management to business value is well thought through, leverages internal strengths, and garners internal support. In short, this approach is aimed at steering a company toward the optimal pathway.

ACKNOWLEDGMENTS

I had a great deal of help in writing this book. I have drawn heavily from the thinking and experience of my colleagues at Arthur D. Little, which has some of the world's leading practitioners in environmental, health, and safety strategy and management.

Ladd Greeno has been a source of encouragement, inspiration and support to me for years, as well as being a seminal thinker and practitioner in the field.

Rob Shelton's thinking had a major impact throughout the book and contributed greatly to the ideas associated with Design for Environment, Industrial Darwinism, value chain analysis and the "green wall."

John Willson's work in shared services, environmental productivity and assessment of environmental management processes was a valuable source of insight.

Michael Isenberg researched many of the examples used in this book and was the principal driving force behind the creation of Chapter 6.

Kate Fish, Director of Sustainable Development at Monsanto Company, generously shared her insights into the lessons learned from Monsanto's implementation of a bold new environmental business strategy.

Jonathan Shopley made many helpful suggestions on the manuscript.

I could not have prepared this book without Patricia Mahon. Every word owes its existence as much to her as to me. She is the finest editor I have ever worked with.

I am grateful to Bill Christopher, the editor of this series, for his patience, persistence, understanding and for originating the idea that led to this book.

Thanks, Maryanne, for being my strongest supporter. You endured the evenings and weekends with patience and even found time to improve the manuscript. Most of all, thanks for sharing the vision.

I.

Business Value and Environmental Economics

THE ULTIMATE CHALLENGE for today's environmental manager is to deliver both environmental performance and business value. But what do we mean by business value? And what is its link to environmental management?

There are many ways to slice this "value" cake. Providing shareholder value means creating earnings growth and share-price appreciation. Providing customer value could mean delivering less for less, the same for less, more for less, or more for more, depending on the company's strengths. Another way to think about value is to envision a numerator and a denominator: the numerator represents net income and the denominator represents the sum of investment, net assets, capital, and head count. Value comes from maximizing the numerator and minimizing the denominator.

In the end, however, it's stakeholder value that may figure highest in achieving business value. Ask the question: "What is important to our customers, shareholders, employees, and the communities in which we operate?" The response will help determine the right strategic objectives for the company, such as growth or innovation, and the right objectives for operational effectiveness, such as lower cost, faster time to market, or better safety. Achieving these objectives provides business value.

A purely financial definition of value has the appeal of simplicity. But it lacks the depth of a more robust definition that shows value in terms of stakeholder satisfaction. Such a definition goes beyond current financial performance and includes the strategic and operational achievements that drive future financial performance. In this book, we will start out with a simpler definition of business value and then progress to a more robust definition. The simpler definition states that business value comes from minimizing cost and maximizing benefit. This cost/benefit relationship is at the heart of environmental economics and is the foundation upon which the robust definition is built.

The Tragedy of the Commons

A fundamental concept that can help us understand today's environmental economics has been called "The Tragedy of the Commons."[1] This concept has been eloquently captured by ecologist Garrett Hardin, based on the work of English economist William Forster Lloyd during the 1830s.[2] Lloyd asked the question: "Why are the cattle on a common so puny and stunted?" To understand the response to this question, Hardin asks that we imagine a pasture open to all, a so-called "commons." As a herdsman using the commons seeks to maximize his gain, he implicitly asks "How will it help me if I add one more animal to my herd?" According to Hardin, the answer has two components:[3]

1. The positive component is a function of the addition of one animal. Since the herdsman receives all the proceeds from the sale of the additional animal, the positive utility is nearly +1.
2. The negative component is a function of the extra grazing created by one more animal. Since, however, the effects of overgrazing are shared by all the herdsmen, the negative utility for any particular decision-making herdsman is only a fraction of –1.

Once the two components are added together, the conclusion for the herdsman is obvious–add another animal to his herd. The full gain is his ("privatized"), while the cost is shared ("commonized"). Since this is the conclusion reached by every herdsman using the commons, the result is that all herds increase until the carrying capacity of the commons is exceeded. Slowly but surely, each animal has less and less to eat until there is not enough to subsist on. Moreover, knowledge of the situation on the part of all the herdsmen won't change a thing. Do you want to be the one to bear each incremental cost while realizing none of the gain every time a fellow herdsman adds to his herd? Do you trust others to do the right thing? It takes just one defection to lead all to conclude it's better to gain while there's still something to gain. There's your tragedy.

Hardin argues that there are only two effective ways to escape the tragedy of the commons. One way is to sell off the commons as private property. This privatizes the cost as well as the gain from using the property, creating an inherent incentive to use the property wisely. The other way is to keep the property public, but regulate the right to use it. Typically, the regulation carries with it some form of penalty or coercion associated with misuse of the property, and has an effect similar to privatizing the cost.

Environmental Economics

What does that have to do with environmental economics? Much of environmental management is driven by the need to address the effects of pollution, which is essentially another form of the tragedy of the commons. Instead of taking something out of the commons, we are putting something in: waste. The polluter asks, "What is the benefit to me of preventing my pollution?" He finds that his share of the cost of pollution–the incremental effect on his health or enjoyment of the environment–is far less than the cost of preventing the pollution. The benefit of not preventing the pollution is privatized, and the cost is commonized.

In the absence of employing one of Hardin's ways out of the tragedy of the commons, we find ourselves locked into a system of creating more and more pollution until the carrying capacity of the earth is exceeded. Just look at the environmental havoc created in Eastern Europe and in large parts of the developing world and see how rational human beings, pursuing their self interests, have polluted their local environments.

Thus, coercion in the form of regulation is applied to make the cost of pollution to the polluter at least equal to the benefit. However, the reality of regulation is that it is not practical to equate cost with benefit. Regulators cannot easily quantify the benefit to the polluter of each decision to pollute, nor can they readily enforce the penalty. In the end, costs from regulatory penalties bear little relationship to either the benefit to the polluter or the total societal cost of the pollution. As a result, the extent to which these costs influence the actions of polluters will depend upon the risk tolerance of the polluters, their assessment of the likelihood of getting caught, and any prior experience with the regulators.

The regulatory costs we have been discussing are the direct costs imposed by the regulators. Another category of costs is the indirect consequence of regulatory coercion, such as the loss of revenues because of the inability to operate production equipment that does not have an environmental permit.

Even if we assume that the direct and indirect costs imposed by regulation are equal to or exceed the benefit of pollution, this does not fully address the problem of pollution. The simple fact is that most environmental regulations do not seek to eliminate pollution, they merely seek to limit it. As long as that is the case, some aspect of the pollution benefit will exceed the regulatory cost.

Are there sources of business value associated with environmental management beyond avoidance of regulatory cost? We have all heard arguments like these: pollution means waste, waste means inefficiency, and inefficiency is bad for business. The recent lore of "pollution prevention" is rife with examples of manufacturers who realized dramatic raw material and energy cost savings, yield improvements, and sometimes commercial value in their waste, versus modest capital and operating investments. Where it is clear that the costs of pollution exceed the benefits of not preventing the pollution, the interests of business and environmental protection are aligned. But it is not always clear that the costs exceed the benefits. It is often hard to quantify the costs and benefits. And some wastes simply do not represent lost product or raw material to the manufacturers that generate them. Extracting value from such wastes must await a more sophisticated system of industrial ecology–a dovetailed industrial infrastructure of enterprises that work together like a natural eco-system. The waste from one entity serves as the raw material for another.[4]

Business Value

It is not necessary to rely solely on the resource-efficiency argument to make the case that good environmental management can provide business value. All that is needed is a more expansive, robust, understanding of the full set of costs borne by polluters (see Figure 1). Earlier, the polluter's costs were described as the amount of coercion or

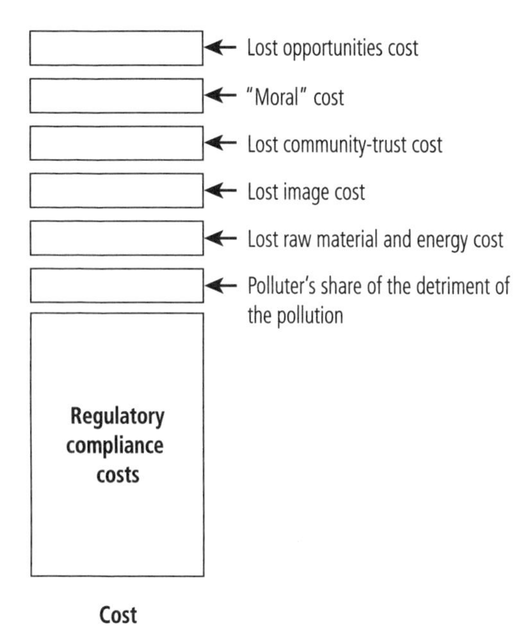

Figure 1. Full costs borne by polluters

tax imposed by regulators. Here, based on the assumption that the amount of the coercion or tax is sufficient to have the polluter deal with the pollution (at least the regulated component of the pollution), the benefits of the pollution are "zeroed out" with the regulatory compliance costs. These include pollution-control equipment capital and operating costs, waste disposal fees, training costs, and record-keeping costs. However, we know that regulation seeks to limit pollution, not to eliminate it. To the polluter, there may still be some benefit to not dealing with unregulated pollution. Stacked up against this benefit is a suite of costs (those shown in Figure 1). Included in these costs is the polluter's share of the detriment of the pollution and loss of some investment made in raw materials and energy. On top of these are costs that include a tarnished image in the eyes of customers and employees–lost sales and lower employee morale; community concerns about chemical exposures, which could diminish the ability to operate; and the cost of lost competitive opportunities. Superior environmental performance or image can lead to increased sales or market access. The ability to obtain environmental permits faster than the competition allows one to get new products to market faster. Failing to capitalize on these opportunities carries with it a cost. There is also the intangible cost of doing something that may be regarded by employees, customers, communities, and other stakeholders as morally improper.

Some of these costs are extremely difficult to quantify, while others such as regulatory compliance costs are easier to quantify but often not fully captured. Since cost efficiency is an obvious source of business value, there has been a recent emphasis on improving the tools available to capture environmental costs.

As awareness grows that these costs are real and "private" rather than "common," so does the attention to managing them properly. This is the real lesson of the tragedy of the commons. William Forster Lloyd not only observed that the fields of the common were bare, but that the privately owned fields adjoining them were fertile.[5] Those who bear the full costs see value in managing them.

Some companies have engaged in an environmental cost accounting analysis only to discover that the environmental costs were an order of magnitude higher than initially supposed. However, accounting for environmental costs was only half the battle. A better understanding of environmental costs has helped these companies reduce their operating costs, segment their product costs, make more informed pricing and product decisions, and select cost-effective raw materials.

Improved cost management is at the heart of the business value that environmental management can deliver. In addition to the suite of costs that range from regulatory compliance costs to the cost of failing to capitalize on competitive opportunities, there is another aspect to consider. Every type of environmental cost faced by a company is also faced by its customers. There is revenue to be had by reducing the costs that your customers bear. "Green" marketing has not had an unblemished history of success, but there is no doubt that certain products and services have a competitive advantage by virtue of their environmental characteristics.

Water-based solvents and paints can lead to fewer health problems during use and are less costly to dispose of. A product take-back program saves a customer's disposal costs and allows for the recycling of components that have some value to the manufacturer. And, yes, cosmetics that have not been tested on animals address the moral cost some people associate with using traditional cosmetics.

Green Products

Throughout the last decade, companies began producing hundreds of new "environmental products," from disposable diapers to laundry detergents. But which products are profitable and still on the market today? The ones that deliver long-term value to consumers.

In order to respond to shortage of landfill space in the United States in the 1980s, many of the major consumer-products companies designed biodegradable trash bags. However, the products did not biodegrade once buried in a landfill and were of lesser quality than other competitive products. Thus, the companies had created new products that were supposed to fill an environmental niche but didn't incorporate basic customer needs.

Companies should use an environmental lens as consumer-products companies did, but that lens should focus on environmental concerns that lead to product improvements. For example, customers demand products that are smarter, faster, or cheaper. Consumers have always been interested in quality and cost advantages, and environmental issues can lead to enhanced customer satisfaction.

To assert cost leadership, Konica is marketing a dry photo-finishing process that will eliminate the need for hazardous chemicals, while reducing the cost of developing photographs. To increase production, General Electric is reducing the solvent content in its high-solid, pressure-sensitive adhesives, thus allowing the company to expand its output before reaching regulatory limits for volatile organic compounds.

Consumers respond to environmentally-based product advantages. Ultraviolet (UV) light and electron beam (EB) inks, both water- or vegetable-oil based inks, are less toxic and easier to dispose of than petroleum-based inks. While overall growth of the ink market has averaged 5.6% over the last five years, UV and EB are growing at 13% and 8%, respectively–the fastest growing sector of the market.[6]

The message is clear. Industrial customers and individual consumers all want better products. Customer-driven needs encourage and drive successful product innovation. Taking advantage of these new opportunities requires applying an environmental lens to better understand and satisfy consumer demands.

Thus, the pathway for realizing the business value of environmental management involves developing a strategy that will improve the management of the suite of costs caused by the company's environmental impact and realize the competitive opportunities around helping customers address their costs. The next chapter explores the first step in developing such a strategy.

II.

Envision the Future

AN ORGANIZATION'S STRATEGY PROCESS begins with vision, an image of a desired future. Vision answers the question, "What do we want to create?" It describes where we want to go and what we will be like when we get there. It embodies what is important to people in the organization, and gives meaning to being part of the organization. Vision is aspirational.

The most powerful visions are shared among a community and elicit enormous creativity and excitement. Vision is essential to building something new and wonderful. A shared vision is a kind of organizational glue that gives coherence to the different things people do, and connects them to something important. Vision is inspirational.

Without vision, significant change is not possible. Sometimes change may happen because of necessity, but change under these circumstances rarely sticks. We have all seen companies whipsawed from one strategic change to another, based on the need to shore up a falling stock price, only to end up worse than they started. Diversification, globalization, decentralization, demerger, business process reengineering, outsourcing, staff reductions, and on and on. For most companies these are pathways to nowhere because they try only to address today's problems, fixing or maintaining something old or broken. In contrast, tomorrow's opportunities can be reached only by those who are excited about building something new and great.

The link between achieving wonderful environmental results while promoting the core business purpose of the company may be the basis of the most powerful and creative vision of them all. Why? Because it has great potential for aligning individual employees' values with the overall corporate mission. As individuals we may take different paths to satisfy our spiritual and motivational needs, but most of us have some strong emotional connection with protecting the environment. It binds us together. At the same time, we choose to stay with the companies we work at because something in their core purposes aligns with what is important to us personally. Link up these two "shades of green" and tap into an emotional force that will fuel the most extraordinary accomplishments.

Skeptics may say, "I've seen visions come and go at my company and none of them did what you are saying." Granted, but each of these "visions" was probably deeply held by one person at the top of the organization and then crammed down the throats of the rest. No wonder they didn't do anything.

How to Articulate a Vision

An organizational vision that aligns with the personal visions of individuals in the organization has the best chance of being realized. Though there is no one way to arrive at a shared environmental vision that's right for a particular company, what's critical is that the vision is shared by, not with, individuals in the organization. Strive to arrive at an environmental vision in a collaborative, inclusive, bottom-up way, not by cascading it down from the executive suite. A top-down vision that is disconnected and unaligned with personal visions will never be realized. The role of leadership should be to ensure that the organizational vision is a collective one.

Assemble a strategy development team with members drawn from interested parts of the organization. The team's first task is to articulate and clarify the vision. The team members should be encouraged to open their minds and think creatively. To get beyond conventional wisdoms about the future, the team should really stretch:

- Think about the fringes, the extreme possibilities
- Look to the result, not the process of getting there
- Focus on what it wants to achieve, not on what it thinks is possible
- Think about what it wants, not about what it doesn't want

Take the team away from the office for a few days. Envision (in the present tense) the company the team wants to create (say, in five years). Set the stage by assuming that the organization and team have been successful. Use the following prompts as a starting point for fleshing out the vision:

- Who are the stakeholders of the organization we have created? What do we do for them? How do we create value for them? What is the secret of our success? How do we measure success?
- What major results have we accomplished?
- What is it like to work in this organization and on this team?
- What are the core values?
- What is our reputation in the marketplace? How do we differentiate ourselves?
- What are our products and services? How have they changed over time?
- What have we learned over these five years?
- What is our unique contribution to the community and world around us? What is the impact of our work?
- What are we most proud of? What have we done that has contributed to our success?

Take the time needed to reflect deeply on these kinds of questions. Share the highlights among the team. Then organize the various elements into a first-pass vision for the organization. Categories such as quantitative results, contribution to society, values, reputation, work environment, and satisfaction to stakeholders can help make sense of the mass of input that has been generated. Be creative. Use visionary language to create the vision. Don't launch too early into word-smithing. Are the results inspirational? Aspirational? Meaningful? Believable? A truly successful organizational vision will be one that individuals are proud to call their own.

With a shared vision in hand, and the energy and emotion level of the team at a high point, take stock of how the vision compares to your company's current situation.

III.

Understand the Current Situation

The next step in defining a successful strategy is to assess the company's current situation in light of its long-term vision and appreciate the gap between the two. An essential element of this gap is the tension that arises in understanding how far the organization is from achieving its vision. But this tension is a positive and creative force that draws everyone toward the vision. The better we define our vision and understand our current situation, the more clearly we see and can be positively influenced by the gap.

A strategy can be defined as the choices an organization makes to achieve its vision (see Figure 2). The main thrust of a vision often involves satisfying or benefiting a company's stakeholders. It follows that a company's strategic choices flow from a desire to satisfy the needs of its stakeholders, taking into account its relative competitive position. A company typically has four primary stakeholders–customers, employees, owners, and the communities

Figure 2. Role of strategy in achieving the vision

in which it operates. Each stakeholder has needs that can be satisfied in different ways. And each company has competitive strengths and weaknesses in how it can satisfy the needs of its stakeholders. The strategy a company selects is distinctively guided by the desire to meet the needs of its stakeholders by accentuating its strengths and eliminating its weaknesses. The company sets its business objectives, ensures that its processes are designed to achieve those objectives, and then aligns its organization and resources to carry out the processes. A good understanding of the current situation is based on an assessment of all these aspects of strategy.

The team should examine the company's markets, competitors and current business strategy so that it can build a solid understanding of the company's business situation.

Aspects to examine include:

- Overall business trends, such as increased acquisitions or divestitures, increased global presence, or organizational innovations
- Plans for new products and processes, potential increases in manufacturing/production, and expected research investments
- Competitive forces, such as time to market, cost management and quality
- Changing customer and market needs
- Operational innovations, including applications of information technology
- Competitor actions

Also look carefully at internal and external issues affecting the company–both today's issues and those bubbling up for tomorrow. (We will discuss *issues analysis* in some detail in Chapter 4 as we explore scenario planning.)

Then assess the following three areas (looking for both strengths and weaknesses):

- Alignment of environmental initiatives with the overall business direction
- Processes in place to understand if the company is "doing the right things right" and also to understand company strengths
- Company resources and organization

Assess Alignment with Business Objectives

The extent to which individual and group goals align with the overall corporate vision determines whether the organization succeeds or fails. The challenge for those involved in environmental management is to ensure that their environmental vision aligns with the organization's overall strategic vision–and also in how the organization chooses to achieve that vision. It is critical to assess the extent to which environmental objectives are aligned with business objectives (see Figure 3).

Consider the electronics industry, where the ability to get new products to market quickly is the difference between success and failure. In the case of Intel, for example, there's a clear link between business success and the ability to obtain environmental permits faster.

And consider the chemical distribution industry, where the effectiveness of customer service can determine

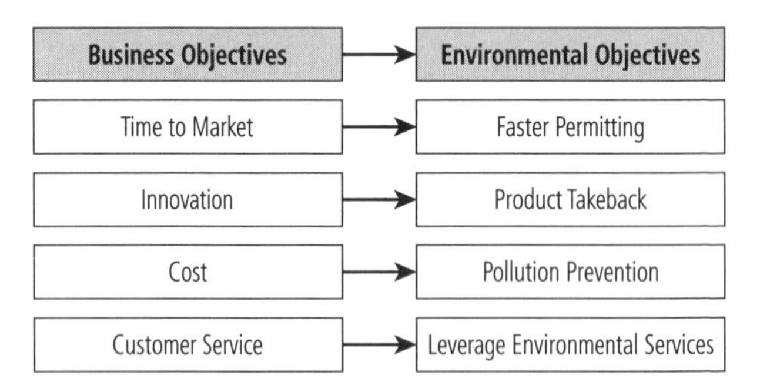

Figure 3. Aligning environmental objectives with business objectives

winners and losers. Ashland Chemical bundles its environmental capabilities with products and services sold to smaller customers that lack environmental resources. This furthers Ashland's overall business objective to provide unique and responsive customer service. Again, the link of environmental capabilities and business value is clear.

But what if it is not clear that the company's environmental objectives link to its business objectives? The "value chain" is a particularly valuable tool to find the links.

Use the Value Chain

The value chain can be used to analyze each value-adding business activity, from initial design to end of life, and find ways for environmental management to contribute to business objectives (see Figure 4). The value chain is helpful because it:

- Identifies the direct linkages between current environmental objectives and business objectives
- Provides a structured way of identifying gaps and new value-creating environmental activities that can fill these gaps. This may be particularly true for those links in the value chain unrelated to manufacturing.
- Is a standard part of most business managers' vocabulary. Once customized to the specific situation of the company, the value chain becomes a powerful way of communicating the value of environmental activities to even the most hard-boiled business person.

Environmental/Business Value Chain Analysis

Business Objectives	Design/ Investment	Raw Materials Acquisition	Manufacturing	Distribution	Sales & Service	End of Life
Product Quality						
Cost				Reduce packaging costs to meet customer desire for low prices		
Time to Market	Streamline permitting process to reduce time required to build or modify facilities					
Customer Focus/Service						

Figure 4. Environmental/Business value chain analysis

Develop a Value Portfolio

Once the environmental activities that are providing business value have been identified, they can be arrayed on a "value portfolio" (see Figure 5). Arthur D. Little developed this value portfolio to help organizations see any potential imbalances in their set of value-adding activities.

Environmental professionals and many business managers are quite familiar with the left side of this matrix. Regulatory compliance, a safe work environment, and protection of the "license to operate" are threshold activities. If we don't have them, it's like trying to drive

	Environmental Value	Business Value
Future Focus	Beyond Compliance • Risk Management • Regulatory Influence	Strategic Objectives • Growth • Image • Innovation • Differentiation
Present Focus	Compliance • Regulatory Compliance • Safety • "License to Operate"	Operational Effectiveness • Time to Market • Customer Service • Cost • Quality

Figure 5. The value portfolio

a car without four wheels. But the fact that a car has four wheels isn't enough to sell it. Threshold value is value without a real competitive difference.

By looking to future-focused environmental value, we can see opportunities for higher returns than threshold activities. Going beyond compliance by taking steps to minimize risk and influence regulations can pay off in the future. Fewer and less severe bad things will happen–such as spills, explosions and toxic releases. However, unless these successes are linked directly to business value, they are perceived as providing value only to the environmental arena. Future-focused environmental investments are treated as optional–a luxury to be indulged in when times are good, and an easy target for cut-backs when times get hard.

By turning our attention to the right side of the matrix in Figure 5, we can see how environmental activities can be linked to business value. Many environmental professionals have made great strides in enhancing the effectiveness of business operations by finding ways to help improve time to market, customer service, cost, and quality–important attributes that are always on the minds of business managers.

At some point, however, we will reach the limits of incrementalism. Operational effectiveness is not enough to provide tomorrow's competitive difference. When a company gets smaller without getting better, it surrenders today's business. When a company gets better without getting different, it surrenders tomorrow's business.

As companies have started to recognize this, we have seen an emphasis on growth and innovation as strategic objectives. Finding ways for environmental management to contribute to achieving these strategic objectives is essentially hitting a *business value* home run.

For example, savvy oil companies are able to obtain choice exploration and production rights by pointing to their superior environmental performance. Certain electronics manufacturers have landed large delivery contracts by virtue of their product take-back programs. And railroad companies have boosted their market share by showing a better track record for safely transporting chemicals.

It is increasingly important to play in all boxes of the value matrix. They don't all have to get equal emphasis, but there should be some balance to the portfolio. What if the value portfolio is unbalanced? Focusing just on the bottom part, the present, concedes the future to others. Emphasizing just the top part of the matrix has its own perils. A company that is too future focused fails to develop the operational skills needed to implement its aspirations.

Now consider the left side of the matrix: environmental value. An overemphasis here ensures that environmental management remains disconnected from the business and is never able to deliver the full value that should be within its grasp.

Similarly, there is a problem in putting all the eggs in the right-hand side of the basket. In the rush to link all environmental activity to business value, the company may take its threshold environmental activities for granted and create a soft underbelly. All the competitive advantage created can be sucked down the drain by one significant environmental incident or compliance violation. So, look for a healthy balance of focus in the value portfolio.

Assess Processes

Even if a full range of environmental processes delivering business value are in place, they may not be as efficient as possible nor fully integrated with business processes. In quality jargon, this would be "doing the right things xwrong." It may be helpful to examine environmental management processes to ensure they are structured and functioning optimally.

Environmental management processes should be systemic (see Figure 6). They should flow in a logical sequence. First, assess the situation, then plan what needs to be done, then implement the plans, and, finally, review the results to ensure things were done effectively. The systemic aspect comes from including a learning component–taking the information developed during the "reviewing" part of the management system and feeding it back to the

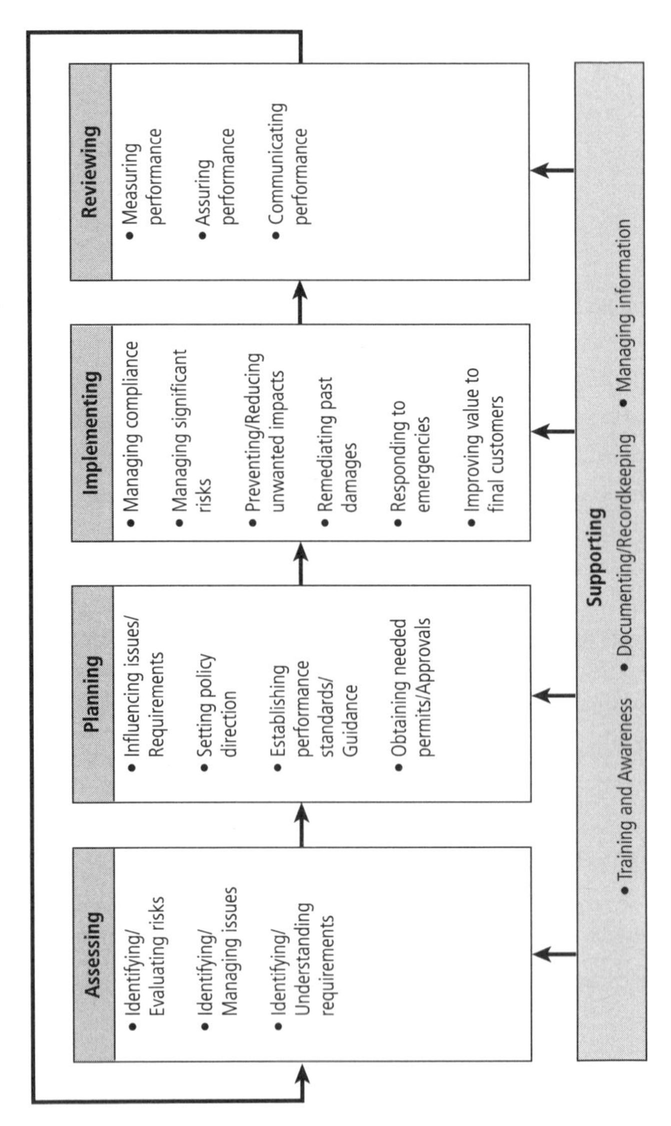

Figure 6. Environmental management processes

start of the system in order to reassess the company's situation in light of the new information. Then plan again, implement, and so on. An assessment of the strengths and weaknesses of the organization's environmental management processes would examine each element by criteria such as efficiency, degree of integration with the business, consistency, continuous improvement, and flexibility (see Figure 7).

Managing Compliance

Management Process Assessment and Rating

	①	②	③	④	⑤
Process Attribute	Needs Improvement		Meets Threshold Expectations		Significant Strength
Involvement of the Line Organization	EHS is staff driven; little sense of line accountability for results.		EHS is staff led, but some line ownership of results exists.		Line-driven, line ownership and accountability for results.
Consistency/ Reliability	Uneven, unreliable compliance.		Mainly in compliance; exceptions are persistent but minor.		Compliance is routinely achieved; strong systems are in place.
Flexibility/Handling of Exceptions	Difficult to distinguish potentially significant compliance issues.		Potentially significant compliance issues are highlighted.		Potentially significant compliance issues receive appropriately expedited handling, resolution.
Cost Efficiency	Heavy EHS staff use, process is not well defined; every time is nonroutine.		Basic process defined; still involves considerable drain on EHS staff time.		Process streamlined; operations staff is well-leveraged.
Continuous Improvement/ Learning	Persistent repeat findings; little evidence of improved compliance.		Fewer repeat findings over time.		Corrective actions address root causes; little evidence of repeat findings.
					Overall Rating:

Figure 7. Environmental management processes evaluation criteria

Assess Resources and Organization

The most elegant management processes will break down if a company's resources and organization are not appropriately aligned to carry them out. An assessment of where one stands should involve more than benchmarking the quantity of resources one devotes. Consider the following factors to be evaluated:

Resources

- Sufficient number of qualified environmental staff
- Environmental staff distribution consistent with business organizational structure and risk profile
- Availability of needed financial and technological resources
- Motivation and incentives for environmental improvement
- Degree of internal customer satisfaction with internal environmental services
- Extent to which line organization is leveraged to carry out environmental tasks

Organization

- Visible top management environmental commitment and support
- Line management accountability for environmental results

- Appropriately high environmental reporting level
- Clear environmental roles and responsibilities
- Culture promoting continuous environmental improvement
- Integration of environmental management into the fabric of the organization

Prepare for the Next Step

Now take the elements of the current situation and identify any patterns and interactions that may exist between them that may affect reaching the company vision. There may be signs of environmental productivity problems (see Figure 8).[7] Identify elements that are outside the direct control of environmental management, but that can be influenced. Look for areas of strength that can be leveraged. Take stock of the extent to which members of the strategy team feel the creative tension that will help move the company toward its desired outcomes. The next step, explored in Chapter 4, is to develop some strategic options.

Process Myopia–Links between environmental business processes are poorly understood (e.g., across functions and product life cycles)

Goal Gap–No articulation of long-term environmental goals, targets

Role Confusions–No clear distinctions among corporate, division and facility environmental roles (e.g., too many people reviewing and interpreting environmental regulations)

Line Dance–No sense of environmental ownership by line managers

Leverage Losses–No pushing down of environmental responsibilities to other functional specialists (e.g., chemicals management, waste vendor audits, acquisition reviews)

Paper Mountains–Strong focus on documentation, managing paperwork–with little understanding of why

Missing Measurements–No ability to quantify environmental results, program accomplishments

Departmentalized Costs–Environmental costs accounted for by department, not process; unrelated to results measures

Blame Game–Difficult to hold line mangers and environmental staff accountable for less-than-stellar performance

Quality Shuffle–Little ability to demonstrate incremental improvement in environmental performance over time

Figure 8. ***Ten telltale signs of environmental management productivity problems***

IV.

Define a Strategy

THE BASIC PROCESS for developing strategic options and selecting an environmental strategy is depicted in Figure 9. It includes these key points:

- Examine the main strategic inputs, including the vision, drivers and scenarios. The principal elements of these are extracted from the situation assessment.
- Use the inputs to develop strategic options for consideration.
- Screen the options for their relationship to business value, alignment with the business, and fit with organizational strengths.
- Select a strategy.

This should be a creative effort. If trade-offs between promoting creativity and maintaining faithfulness to the process need to be made, opt for creativity. There will be greater team enthusiasm and buy-in and richer results.

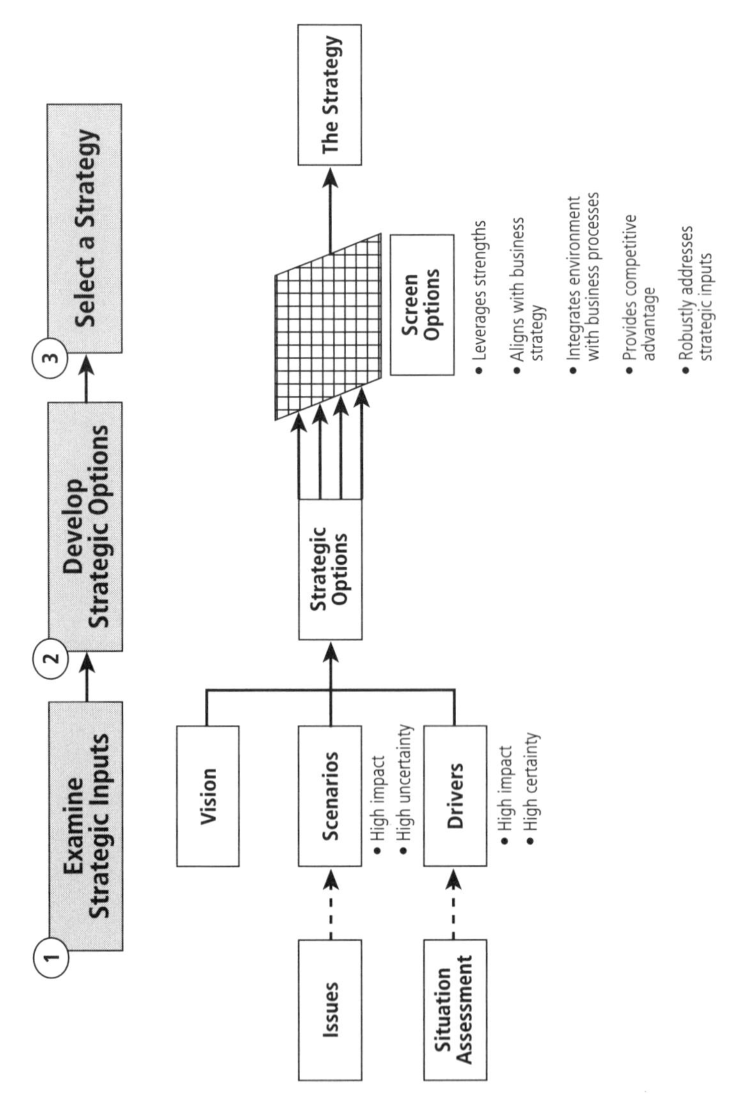

Figure 9. Strategy development process

Examine Strategic Inputs

A robust strategy has several facets. A strategy should be:

- **Aspirational**–serving as the pathway for realizing the vision of the organization
- **Practical**–leveraging the organization's strengths and taking into account forces and events that will likely influence the firm
- **Flexible**–serving the firm well in any of several plausible futures

The raw material for shaping each of these facets is derived from three strategic inputs: vision, drivers, and scenarios.

Examine the Vision

Take a fresh look at the vision articulated during the process described in Chapter 2. Is there an element of providing business value embodied in the vision? Is there an element of making a contribution to society? Is it more operational? Whatever the answers, maintain the integrity of the vision throughout the strategy-development process. Vision is the guiding light.

Examine the Drivers

The bulk of the situation assessment should boil down to a few key drivers–the high-impact influences that will almost certainly come to pass. These include the main business trends, competitive situation (in a relatively stable industry), stakeholder threshold expectations and strengths of the firm–the subject matter of traditional strategy development. However, high-certainty influences will surprise no one, least of all competitors. Thus, it is critical to look to the future and identify high-impact influences about which there may be a high degree of *uncertainty*. If these come to pass and the company is prepared for them, it gains a competitive edge.

Examine the Scenarios

An important tool for looking to the future is scenario planning (see Figure 10). Paul Valéry, the French philosopher, said that "The trouble with our time is that the future is not what it used to be." Scenario planning helps us consider alternative plausible futures that go beyond extrapolating from the present. These scenarios are stories about the way the world might turn out tomorrow. But they also become agents for change as people

Strategy Step	Traditional Approach	Scenario–Based Strategy
Anticipate the Future	Extrapolates from history	Considers alternative plausible futures
Formulate Strategic Objectives	Assumes incremental change from present situation	Uses scenarios to define core objectives
Develop Strategic Options	Promotes variation around a single theme	Uses creativity to define options consistant with scenarios
Formulate Strategic Course of Action	Pursues generic strategy based on existing rules of competition	Defines path and then adjusts later

Figure 10. Scenario planning enriches strategy development

understand their full ramifications. Scenario thinking offers a way to engage senior leadership in challenging familiar assumptions and helping their organizations prepare for, and prevail in, a future in which the rules of competition have been rewritten.

The name "scenario" comes from the theatrical term for the script of a play. The theater audience watches actors on a stage, but reacts as if the story is real. This is called "the willing suspension of disbelief." A good scenario does the same. You know it's a story, but see how it might happen and better understand what you might do as a result. At a minimum, the scenario process will help challenge "mental models," those deeply held images and assumptions we carry in our minds about ourselves, other people, institutions–in short, how we see every aspect of the world.[8] Like a pane of old glass, they frame and slightly distort what we see. Mental models are applied to environmental management all the time. The plant manager thinks "Environmental spending is simply a drain on limited resources." The senior manager thinks "Environmental management is just a compliance paperwork exercise; a waste of time." How do we move to a mental model where environmental management is seen as providing business value? Mental models are slow to change, but they can be changed. It often takes a fresh, bold point of view to make people engage their imaginations and see the world differently.

A classic example of successful scenario planning is how Royal Dutch Shell came up with "higher oil prices" as one of its strategic scenarios in the early 1970s. When the oil crisis hit in 1973, of all the major oil companies, only Shell was prepared for the change. From one of the weaker of the "Seven Sisters," it became one of the largest and most profitable.

Issues are the raw material for scenarios. Although some environmental issues seem to appear abruptly in the national or international consciousness, in reality each issue emerges through a development process. Companies that tap into this process improve their ability to judge the potential effects of developing issues well before they reach the general public or become bases of competition. For each issue, the development process begins with early awareness, followed by scientific corroboration, evolving research and accumulated knowledge, industry participation, opinion leader interest and sometimes a major event that moves the issue into the public arena.

Environmental issues that might affect the company's competitive position need to be identified and their implications assessed along the value chain. Useful tools for identifying issue areas include: life cycle analysis, value chain analysis, stakeholder needs assessment, regulatory assessment, management systems assessment, and environmental audit trend analysis. Moreover, companies can monitor the progress of each issue by expanding company communications and networking mechanisms to include an environmental dimension.

Thinking beyond the present is critical to understanding where threats and opportunities lie. Once the issues are identified, they are categorized by degree of uncertainty and potential level of impact (see Figure 11).

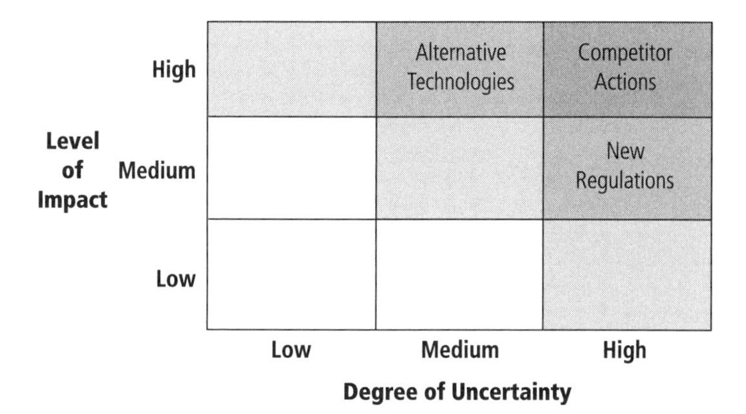

Figure 11. Issues prioritization

Issues typically evolve through four stages of maturity:

- **Latent**–where an issue is in early stages of awareness and there is little support of leadership
- **Emerging**–characterized by available detailed research data, political awareness, and media attention
- **Mature**–where there is established support for dealing with the issue and regulations have been developed
- **Institutionalized**–where it is common practice to manage this issue, processes have been internalized, and change is incremental

Consider questions like these to assess the maturity of an issue: Is there a critical mass of support formed or forming? How much knowledge exists about the issue?

Is the knowledge being communicated? Have solutions been identified and implemented? Who is leading the issue? What is their commitment and influence on the issue? Are there recognizable targets? Are they a focus for anger? Has there been a well publicized trigger event?

As the issues evolve, the degree of strategic freedom for a company to do something about them narrows. So the cornerstones of scenario building are those high-impact issues that are in an early stage of evolution. The greater the degree of uncertainty, the greater the value of scenarios.

The next step is to build a story. The key is to develop an internally consistent view of the future that is sensitive to the threats and opportunities presented by a given set of significant issues. Figure 12 illustrates four stories of the future suggested by two important issues. The process of scenario development would flesh out

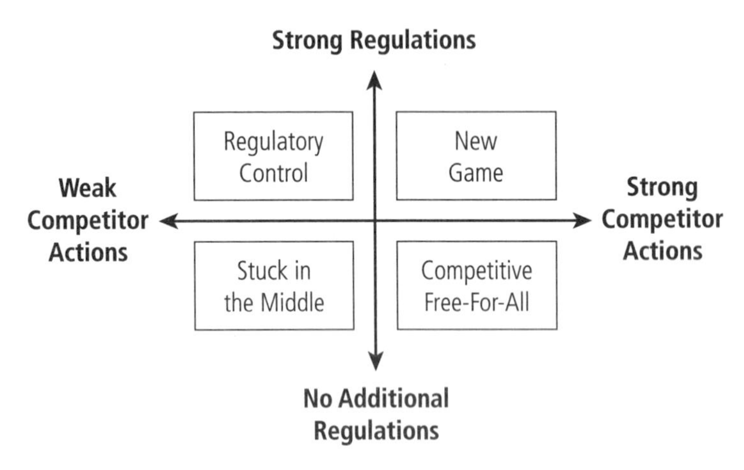

Figure 12. ***Scenarios explore the most uncertain and significant issues***

each of these stories so they could be used as test beds for evaluating alternative strategies.

By considering several plausible scenarios and developing a strategy that is sound for all of them, the company will be prepared to capitalize on whichever future takes place.

Develop Strategic Options

Next, take the three inputs–vision, drivers and scenarios–and come up with strategic options that address these inputs. What might these options look like? Michael Porter, a leading guru of strategy, offers insight on this question. According to Porter, the only way to achieve sustainable advantage is to establish a unique competitive position. Moreover, he says "The essence of strategy is choosing to perform activities differently than rivals do."[9] The implications of these insights are generally the same for environmental management as for other business areas. Thinking back to our Value Portfolio (see Figure 5), we can deliver business value through environmental management in two ways:

- Look to the "different things" that will make a competitive difference in the future, like innovation, growth and product differentiation (the upper right-hand box).
- Look to "doing things differently," either to make the "different things" happen or to make a competitive difference by enhancing operational effectiveness (lower right-hand box).

Either pathway can be a successful route to a winning strategy.

Strategic Objectives: "Doing Different Things"

To illustrate the opportunities for doing different things in environmental management, let's use a metaphor we at Arthur D. Little call "Industrial Darwinism." Just as animals in the natural world employ different strategies to succeed, companies employ different strategies to obtain competitive advantage–including the approaches they choose for environmental management. Figure 13 illustrates the environmental management approaches that might best embody each strategic approach described there.

The lion at the top of the food chain. Like the lion at the top of the food chain, a company can perform to highest environmental standards, and then press to raise the bar for others. Reaching the top first probably fits this company's strengths, but forces competitors to graft a new set of requirements onto systems that were not designed to handle these requirements in the first place.

Here's a current example of "raising the bar." British Petroleum has taken a very strong, visible stance on sustainable development, specifically on global climate change. In the words of John Browne, BP's Group Chief Executive, "The time to consider the policy dimensions of climate change is not when the link between greenhouse gases and climate change is conclusively proven . . . but

	Lion	Hyena	Bird	Rhino
Metaphor	Occupy the top of the food chain	Scavenge food	Live off the host	Hard to kill
Environmental Strategy	Perform to higher environmental standards	Use others' waste as your feedstock	Leverage larger company's environmental capabilities	Unimpeachable environmental performance
Advantage	Raise the bar for others	Lower raw material cost	Lower environmental cost	Avoid environmental vulnerabilities

Figure 13. Industrial Darwinism

when the possibility cannot be discounted and is taken seriously by the society of which we are a part. We in BP have reached that point."[10] The actions BP is taking include: making financial investments to eliminate emissions of volatile organic compounds; setting goals for reducing flaring and venting; improving production efficiency; supporting scientific research through partnering, technology transfer and joint implementation; and generally taking a strong stand on what the oil industry can do to reduce the risks around climate change. By getting out in front early, BP creates a high ground, forcing some competitors to adapt to new requirements they previously denied were important.

The hyena, scavenging opportunities to use others' waste as its feedstock. A company can choose to locate its plant next to or inside the fence of its supplier. Benefits might include lower transportation and raw material costs. This kind of opportunity is generally driven by the needs of two different companies. In an example we know first-hand, a large chemical producer was generating multiple wastes, including isobutyl ester by-products from manufacturing its chemical intermediates. The mixed wastes were treated in a combined wastewater treatment facility that required expansion to meet new emissions regulations. At the same time, companies in the pulp and paper industry wanted to replace methylene chloride with a less-polluting solvent. Isobutyl esters were identified as a viable replacement. Consequently, the chemical producer developed an approach to segregate and purify these wastes, now sells its converted waste (i.e., purified isobutyl esters) to pulp and paper companies, and saved the cost of building a new treatment facility. The needs of two different kinds of companies created the catalyst for making this technology connection. Though "scavenging" may be considered an indecorous word for describing this kind of positive outcome, the point is that these companies were looking much further than their own facilities or industries to find business value.

The bird that sits on the backs of big animals and feeds on the insects that find their way there. Some companies rely on a larger company to do the heavy work–such as developing environmental management systems–and then adopt their practices. Or, a smaller company might turn the environmental performance expectations of its customers to full advantage. Certainly this has been the case in the automotive industry for many years, where the major automotive manufacturers have established performance criteria for their suppliers, including quality, cost-effectiveness, just-in-time methods, safety and environmental issues. For the suppliers, meeting these expectations helped increase their learning and also positioned them as preferred partners to the majors, ahead of competitors.

The rhinoceros, with armor so thick that it can withstand just about any assault. A company can build its protection by having unimpeachable environmental performance. In companies taking this approach, the environmental commitment typically comes from the top of the organization. Effective environmental management systems make sure things get done right the first time; strong accountability mechanisms help ensure the line organization's incentives are aligned with desired behavior; and state-of-the-art environmental audit programs are in place.

The point is that just as animals compete for food and survival, each one develops a unique competitive strategy and space. There is no single environmental strategy that's right for every company: create a strategy that is different, bold and right for your organization.

Operational Effectiveness: "Doing Things Differently"

The second pathway for creating a winning strategy is to emphasize the lower right-hand box of the Value Portfolio (see Figure 5). The focus is on operational effectiveness by "doing things differently." There are many creative embodiments of this approach.

Some companies are managing their environmental activities differently by implementing a shared services model, essentially creating an internal consulting organization. The shared services approach has been applied to the full range of support services, from legal and accounting to purchasing, human resources, and information systems–and, increasingly, to environmental management. Like an external consultant, this group identifies customer needs, engages in marketing, focuses on operational effectiveness and measures customer satisfaction. Though part of the value comes from eliminating duplication of effort, even more value can result from splitting off the environmental oversight function to a traditional corporate oversight group. This enables the environmental service organization to focus its mission on satisfying the internal customer and, thereby, to raise the level of service.

Some companies have taken other approaches. I've been to a small chemical plant where everyone is expected to be capable of doing pretty much everything. Though the plant's environmental complexity is the same as larger plants, its size caused management to question the value of having hordes of environmental specialists on staff. The solution was to train operators in various environmental specialty areas such as lab safety, to get them certified, and then pay them more for each certification. The plant manages well with a smaller environmental staff. Environmental management is fully integrated into operating roles, and operators get training and career opportunities they wouldn't obtain otherwise.

In each of these examples, the company is addressing some aspect of operational effectiveness important to its business. In developing strategic options, the company focuses on those areas that are important to excel at in today's business environment. Then the company develops some ideas for carrying out environmental processes that will support these competitive areas. Let's take a closer look at three important aspects of operational effectiveness to see how environmental processes might be designed to deliver value: time to market, customer service, and cost.

Time to market. If time to market determines winners and losers, changing an environmental process to remove it as the bottleneck can deliver enormous business value. For example, the competitive battles within the computer and electronics industry can be won or lost depending on how fast new products reach the marketplace. The average product life cycle for the industry is 18 months. But the need to obtain environmental permits can produce major backlogs in the ability to get new products to market. Intel, the world's largest computer chip manufacturer, established cooperative partnerships with environmental agencies to streamline the permitting process and create more flexibility. As a result, it is achieving a distinct competitive advantage in preventing delayed plant startups or manufacturing interruptions, improving its market share, and achieving a higher return on assets.[11]

Customer service. In some industries, customer service is the competitive key to selling products. A manufacturer can obtain a significant advantage over its competitors by providing its customers a service that reduces the environmental costs associated with their products. Baxter International, an international manufacturer of medical products, utilizes its environmental expertise to solve its customers' recycling and pollution-prevention problems and, at the same time, reduces its own cost structure.[12] Here are two examples of Baxter's success:

- Baxter has formed an alliance with Waste Management, called Access, to provide customers with technical and regulatory advice. More than 300 major U.S. hospitals have used the service since its inception in 1989.[12]
- Baxter worked with Albany Medical Center to examine the center's use of Baxter's surgical pack. It then redesigned the product to make it 25% lighter and 18% cheaper. Over a four-year period, Albany Medical saved $800,000 by reducing its waste stream, and Baxter reduced its own raw materials and distribution costs.

Cost. Lower cost is always a competitive advantage. Many companies have come to the realization that running eco-efficient manufacturing processes lowers cost. These processes minimize not only the unwanted outputs, like waste and byproducts, but also the inputs like raw materials, energy and water. Annual return on eco-efficiency investment often exceeds 50%, well in excess of most other capital projects.

Clean it. Control it. Fix it. Throughout the 1970s and 1980s, these were the primary environmental strategies used by companies to comply with environmental regulations. To meet new standards, end-of-pipe technology was employed. Then many companies began asking "why can't we eliminate pollution at the source and save money?" Thus pollution prevention (P2) was born: source reduction aimed at lessening the use of or risk involved from a pollutant or hazardous substance, and chemical substitution to replace toxic substances with less harmful materials. Here are some successful examples:

- 3M implemented one of the first applications of a pollution prevention approach, Pollution Prevention Pays (3P). The program (begun in 1975) created 4,590 employee projects worldwide, prevented 1.5 billion pounds of pollution, and saved $790 million in costs. And, 3M continues to accrue extensive good will and public relations benefits from 3P.
- Carrier Corporation redesigned its air conditioner parts and changed its metal-cutting process. The firm eliminated toxic solvents, saving $1.2 million in production costs and improving the overall quality of the product.
- Whyco Chromium developed an organic compound for coating nuts and bolts that cut the number of coats in half, thereby reducing waste and cutting manufacturing costs by 25%.
- AT&T redesigned a circuit-board cleaning process, resulting in the elimination of ozone-depleting chemicals and an annual savings in cleaning costs of $3 million.

In a similar vein, Xerox extended the reach of its thinking to extract value from its products even after they reached the end of their useful lives. Xerox instituted a process for product take back which included product recovery, disassembly, and remanufacture or recycle. It found there was significant potential residual value in its copiers–value it had been losing when the product was disposed of. Many components had a far longer useful life than the product as a whole. Xerox captured this "lost value" in the form of reusable components once the product design was modified to allow for easier disassembly. This Asset Recycle Management Program and Xerox's overall waste asset management processes have increased customer satisfaction and generated annual cost savings of about $200 million.

As the sophistication of environmental processes to address cost has grown, some companies have also developed tools to improve the quantification of the costs associated with environmental processes, including total cost accounting, eco-accounting and environmental cost accounting. Even more noteworthy, some companies have made progress in capturing the business value of environmental processes. For example, Baxter International is expressing its environmental performance in financial terms in response to skepticism by business managers about environmental efforts. Baxter developed a peer-reviewed methodology that shows environmental costs and savings in a green profit and loss statement, reporting benefits of $15.2 million in 1995 and $19.7 million in 1994. On the savings side, the firm lists everything from recycling income ($5.9 million in 1995) to hazardous waste disposal cost reductions ($600,000) and material cost reductions ($600,000). The cost side includes everything from pollution control operation and maintenance ($5.0 million) to corporate environmental affairs and shared multidivisional costs ($1.5 million). According to Bill Blackburn, Vice President of Environmental Affairs and Chief Environmental Counsel, "Now we find business managers talking about the environmental program in terms of the business benefits it can bring."[13]

Tighter Loops: A Fresh Perspective on Operational Effectiveness

There is a common element to the different aspects of operational effectiveness discussed above. The companies cited as examples of creating business value from environmental contributions to operational effectiveness have, in one fashion or another, formed "tighter loops." They have shifted from thinking linearly–seeing the manufacturing process as taking raw materials, processing them and sending the products off to customers and the waste off to nature–to seeing interrelationships and the way things work in a systemic manner, like an eco-system in which nothing is wasted. Intel "tightened the loop" by shortening the time in the design-manufacture-design cycle. By bringing the environmental management process of its customers "within the loop" of the product life cycle, Baxter eliminated costly inefficiency. The companies that made dramatic progress in pollution prevention took manufacturing processes where waste had been seen as "outside the loop" and brought it back in. As a result, they streamlined their processes to squeeze out more value from their raw materials.

There may be value to the concept of "tighter loops" to spark new ideas as we search for strategic options. Consider the possible variations on how to achieve tighter loops.

Closing the loop. Supply the missing element and thereby make something that had been linear into something more systemic and self-regulating. In management processes, measurement of process effectiveness is used to feed back information to allow for more informed assessment at the front end of the process. Then adjustments can be made. In manufacturing processes, seeing unintended consequences or side-effects (such as waste) as part of the overall process leads to applying to them the same efficiency-optimization thinking that had been applied all along to the product loop. In its rudimentary form, it leads to recycling and extracting value out of waste. Taken to its logical conclusion, this thinking results in closed-loop manufacturing.

Tightening the loop. Streamline or accelerate the process to reduce the time or materials involved. Avenues include energy efficiency, raw material efficiency, pollution prevention, dematerialization and design for environment.

Enriching the loop. Link parallel loops and extend the reach of the loop beyond the process at hand. Environmental management processes are integrated with business processes. The manufacturing process-control loop is extended upstream and downstream beyond the process at hand to allow for more systemic control. On a grander scale, the concept of "industrial ecology" involves bringing within the loop all the elements of the vertical supply chain. This parallels what occurs within an eco-system: the waste from one entity serves as the raw material for another. The waste heat generated by the power plant is used in the manufacturing processes of the nearby oil refinery and pharmaceutical plant. The excess sulfur generated by the oil refinery is used as a feedstock by the sulfuric acid plant. The sludge from the pharmaceutical plant's fermentation vats is used as a fertilizer by local greenhouses and in gardens at the refinery and power plant. And so on.

As a result of thinking about loops in this manner, an interesting phenomenon emerges. We realize that learning is at the heart of the systemic structure of loops. The feedback flowing from measurement is the key to the management process loop. In manufacturing processes, the new understanding of the negative effects of unintended consequences, such as waste, allows for the successful application of design efficiency, process efficiency, and process control techniques. In industrial ecologies, the initial appreciation for the potential interconnectedness of independently owned and operated manufacturing facilities leads to an entirely new way of operating the enterprise.

As we tighten our loops, we employ all the skills and thinking needed to become a learning organization. We engage in systems thinking. We obtain broad organizational involvement. We continuously improve our processes. We work toward a shared vision. We make innovations in our infrastructures.

As a result of pursuing operational effectiveness through tightening our loops, we learn how to become a learning organization. We have not only delivered on the traditional bases of business value, but also set the stage for conquering the next management frontier. It is the organizations that are most skilled at learning that will prosper in the twenty-first century.

Select a Strategy

Possibilities like these can lend real excitement to the task of developing strategic options. Once those options are developed, it is time to select a strategy. Some sort of screening criteria can be used to winnow down the strategic options to select the strategy that is optimal for the company. Screening criteria could include the following:

- How well the existing strengths of the organization are leveraged
- The extent to which the environmental strategy aligns with the overall business strategy
- The extent to which the environmental strategy integrates environmental management with business processes
- The degree of competitive advantage and business value provided
- How robustly the strategic inputs (particularly the scenarios) are addressed

Avoid the temptation to integrate several strategic options into one. While it is advisable to build some flexibility into the strategy, making it overly complex will make it more difficult to communicate and implement. The strategy will have to be implemented by a broad cross-section of the organization. Keep it simple.

The next step is to develop an implementation plan.

V.

Develop a Plan

Having settled on the strategic direction, the next step is to operationalize the strategy. A plan can lay out the design of the management processes needed to realize the strategy. Processes involve the sequence and logic of work that gets done. Since work gets done by people, processes are defined by relationships between people in an organization. Thus, process design must take into account the appropriate organizational elements that should be aligned to the processes.

Experience shows that effective processes are based upon work that is performed across functional boundaries. Only when the right set of diverse talents and perspectives are aligned to achieving a common purpose can true progress be made. But in many organizations, the deepest attachments tend to be formed to functional teams. When an individual's motivation and sense of self worth in the work context are wrapped up in his or her functional role, it is hard to break out of the "silo." This creates challenges in designing new management processes.

Adding to this challenge is the peculiar relationship that often exists between environmental staff and the line organization. There seems to be a kind of "green wall" between them.[14] Business managers may see environmental management as a necessary evil, but its cost and influence should be minimized. Environmental managers assume that the business value of their efforts is obvious. If environmental management is to be fully integrated into the business, this "green wall" must be overcome. In thinking about setting up a plan, let's look at an example of successful environmental integration into a business process to see what we can learn: Design for the Environment.

Design for the Environment

A number of companies, particularly manufacturers of computers and other electronics equipment, have included consideration of environmental issues in the product development process. This integration of environment into a business process, often termed Design for Environment (DfE), has resulted in products that are less costly to manufacture and that better meet customer needs. We can learn a range of implementation lessons from DfE that will help more broadly in considering implementation planning:[15]

- **Include environment early in the product-development process.** This makes it easier to design the product with the environmental characteristics that will differentiate it in the customer's eyes. This also avoids the redesign, rework and delays in getting to market, which would occur if environmental limitations only became known well down the road.

- **Create a cross-functional team.** DfE requires the application of many different talents–R&D, manufacturing, marketing, environment–working together. If one functional component is dominant, the results will be skewed and other members of the team will lose enthusiasm and fail to support the results. DfE should belong to the design team as a whole.
- **Ensure the process is customer driven.** The goal of DfE should be to create products that are more attractive to customers than competitors' products. The business value of differentiation is clear. Early attempts at DfE were overly driven by the desire to create so-called green products, whether or not they had attributes that would be attractive to customers. As a result, they failed.
- **Use pilot projects to build momentum.** Top-down, corporate-driven programs typically encounter stiff organizational resistance. A better way to incorporate DfE into the fabric of an organization is to start small, perhaps with one product line, and make sure the business owns the process. The pilot design team will learn what does and doesn't work within the organization and be in a position to transfer this learning to other design teams. As other parts of the organization see the success of the pilot project, they will develop a desire to create DfE teams of their own. This is how momentum gets created.

- **Develop simple tools and metrics.** Part of the problem with early DfE efforts was the reliance upon "life cycle analysis" to determine the relative "greenness" of the proposed product versus competitors' offerings. Traditional life cycle analysis could not reach definitive conclusions and ended up just slowing down the product-development process. Much simpler tools have been developed that support, rather than inhibit, the process. Similarly, simple metrics are essential to show the effectiveness of the DfE effort. These metrics should be framed with benefits in mind, such as added value, cost savings and competitive advantage–rather than purely environmental results.

Thinking back over these lessons, we recognize that some relate to the design of management processes, particularly in relation to the use of metrics, and others are organizational in nature. Let us take a closer look at these two important aspects of implementation planning.

Effective Use of Metrics

It's hard to stress too strongly the importance of developing the right metrics that show how environmental activities align with and contribute to the company's business objectives. For example, stakeholders' needs might comprise increased sales and market share, competitive operating costs, attractive working environment, and compliance with regulations and best practices. And corresponding business objectives might cover the need to increase production capacity, decrease production cost, improve workplace safety, or improve training retention. Thus, the environmental performance metrics might focus on permit cycle time, raw material and waste, injury severity and frequency, and test scores. By aligning the environmental performance metrics with business objectives, stakeholders can see clearly how their needs are being satisfied by environmental activities.

We will examine three key measurement areas for capturing and understanding the value of different environmental activities: those related to compliance, to beyond compliance, and to operational effectiveness activities (three out of the four boxes of the Value Portfolio). Each requires a unique set of metrics.

Compliance. The business value of a threshold area such as compliance is to achieve the right result as efficiently as possible, spending only what is necessary. If a company is not in compliance, it has to get in compliance. But it also has to know the cost of compliance. Benchmarking is an essential tool to help compare those costs to other companies. If a company finds it is achieving high results, but at higher cost than others, it needs to reduce its costs. If costs are in line with the benchmarked companies, but results are not as good, the company has to improve its results. If both costs and results are out of line, the company has a lot of work to do. And, if it achieves the same high results as others, particularly competitors, at lower cost, it has created business value.

Beyond compliance. A different set of metrics is needed for "beyond compliance" activities. Here, the main thrust is risk management, taking steps today to avert problems in the future. The source of business value is primarily cost avoidance. Some of us may know it's not always easy to make the case to spend now in order to avoid future problems for which we have little prior experience. But others may have a track record from which we can learn. It is possible to quantify risk, scenario by scenario, and come up with a risk inventory.

As shown in Figure 14, the inventory can be mapped by the dollar magnitude of losses, discounted by their likelihood of occurrence. Benchmarking then helps identify opportunities to mitigate risks that are viewed as unacceptable. These opportunities represent the potential returns on risk mitigation investments. Once it's set out in these terms, the actual cost-benefit analysis is fairly straightforward.

Operational effectiveness. For operational effectiveness, we can draw on traditional metrics and analysis, particularly in the area of cost savings. Presumably, there is a track record to draw on concerning costs. The only issue is whether the investment in the cost-saving project will give a better return than other ways the company can spend its capital. So, we use a traditional capital budgeting approach to examine projects that might include pollution prevention, packaging reduction, resource use reduction, workers compensation management or energy efficiency projects. First quantify historical costs, and then quantify

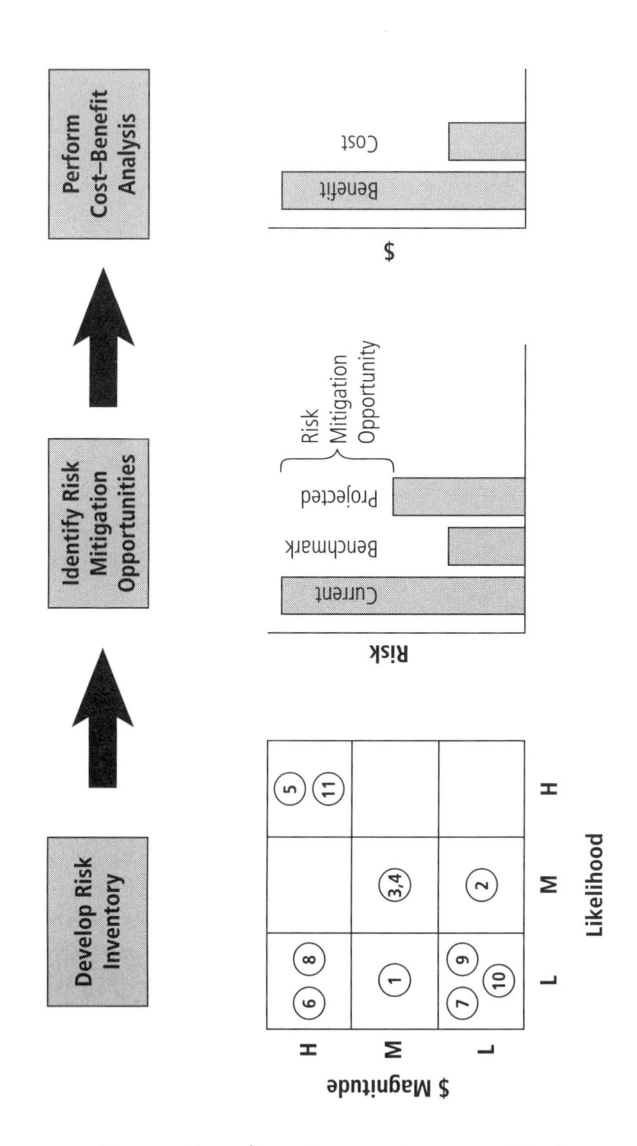

Figure 14. The value-creating potential of "beyond compliance" activities

project investment and cost-savings return. For example, how do these projects reduce costs for waste disposal, workers compensation, raw materials, energy and water? Last, compare the project rate of return with the company's incremental cost of capital. If the project rate of return is better, it ought to be considered side by side with other capital projects the company has in mind. The project's rate of return might even be substantially better than competing projects–some companies have found that their rate of return for pollution prevention projects exceeds 50%.

Organizational Factors

Often the hardest part about integrating environmental considerations into the fabric of the business involves how people relate to one another. Tearing down the "green wall" between environmental and business managers requires the efforts of both parties, though the environmental manager generally has to take the lead.

A first step is to identify those business functions or processes with which it is most important to establish environmental linkages. The Value Chain analysis will be helpful here (refer back to Figure 4). Then assign an integration champion who has a good relationship with and understanding of the targeted business function or process, as well as the relevant environmental issues. The champion's job is to ensure that relationships are built up with key business individuals, relevant environmental information is provided to them, and the value of environmental management is properly communicated. That means turning an environmentally-focused language (typically jargon laced with regulatory acronyms) into language the business culture understands.

The environmental function should also act more like a service business. Structurally, this may involve splitting off the environmental oversight activity from a group dedicated to providing environmental services. This reinforces the message that the environmental function is there to support the business. Thus, the environmental function must do the things that make a service business successful, including:

- **Listening to customers.** Take a systematic approach to understanding the internal customers' business situations and related environmental needs. Periodically agree upon service contracts that specify the expected nature of the services to be provided. Attend staff meetings. Establish focus groups to gather ideas about how to improve the value of environmental services.
- **Developing a services strategy.** Articulate a clear value proposition that differentiates the environmental function from the competition. Then ensure that the function can deliver on this distinctive capability.
- **Marketing the offering.** Create awareness about what the company offers. Give the customer a basis to prefer the function over competitors. In addition to providing a distinctive capability, this means developing personal relationships with the customers and showing genuine interest in their situation.

- **Matching the solution to the need.** Corporate environmental groups are often viewed with distrust because they provide solutions to needs beyond those expressed by the business. The environmental group may become intrigued by the intellectual challenge of delivering a state-of-the-art solution when this may not be what the customer wants or can afford. A service organization is there to serve, and it cannot get close to the business if it does not meet the trust of the business.
- **Learning from customer experience.** Put processes in place to solicit customer feedback on service performance. Integrate the learning from this feedback into the service-delivery process.

Business management also needs to make an effort. While we can't wave a magic wand and have business management take up the initiative when it is not predisposed to do so, we can help increase the likelihood of progress by making business management more knowledgeable and accountable.

Making business managers more *knowledgeable* involves all the efforts of the environmental staff to demonstrate the business value discussed above. It also involves creating opportunities for business managers to learn from themselves. Some companies have established environmental policy councils, composed of senior line organization managers, to help set corporate policy. In practice, they become forums for sharing an understanding of the value of environmental management. The leaders get tips from other leaders and the laggards realize it would be to their advantage to improve. Nobody can better articulate business value than a business manager.

Making business managers more *accountable* involves setting clear performance expectations and individual performance objectives, measuring environmental results, and tying performance reviews and compensation to these results. Accountability mechanisms are often coercive and heavy-handed, but they do get attention. Sometimes, they are just what is needed to escape from a feedback loop that is driving behavior in an unwanted direction.

The discipline of "systems thinking" has taught us that logical, rational people can get caught in illogical, irrational and destructive patterns of behavior.[16] Yet each individual choice and action they take seems logical and rational at the time. It is the ability to see the causal relationships between the choices and actions that helps us resolve this paradox and act more productively.

One company found itself facing a disturbing set of behavioral patterns. Though the number of its environmental compliance problems fluctuated over time, the long-term trend showed an increase. Moreover, the cost to deal with each new problem kept increasing. And the number of environmental lawyers needed to challenge regulatory agency enforcement actions kept growing. These lawyers were particularly successful at going to court to defeat regulatory agency actions. Nobody was thrilled to have the number of lawyers increase, but they feared that without the lawyers they would have more problems and higher costs. The better the lawyers were, the more the company relied upon them. However, the reliance on lawyers to fix the problems reduced the need for people who could prevent the problems from happening at all. Thus, the number of environmental professionals kept decreasing. Perhaps the most telling pattern was the decline of compliance mentality within the line organization. This was troubling, because sooner or later it would result in the type of compliance problems that lawyers couldn't readily fix.

The interrelationship of these patterns is shown in the "systems loop" depicted in Figure 15. The box in the center indicates the symptom of the underlying problem–

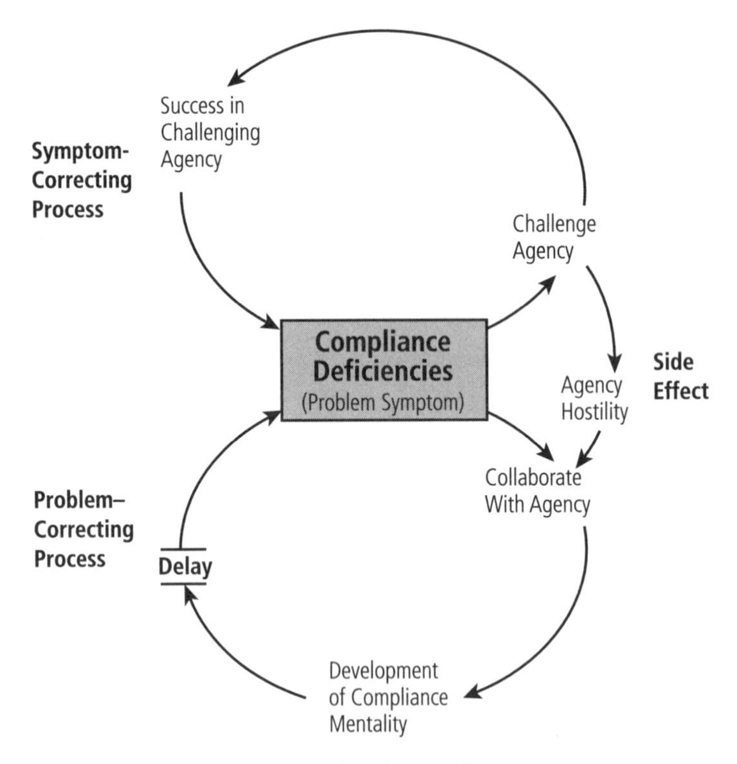

Figure 15. Systems loop

compliance deficiencies. The upper loop shows the process used to correct the symptoms. The lower loop shows the longer, more difficult process used to correct the underlying problem. The arrows indicate that one element influences another. The element off to the side is an unintended consequence of an action–a side effect–that compounds the problem.

The upper loop shows that when a compliance deficiency occurs, the immediate solution is to enlist the lawyers to challenge the regulatory agency. Owing to the skill of the lawyers, the symptom is quickly (though expensively) resolved. The lawyers go to court and defeat the agency. This reinforces the perception there is no way to deal with the problem other than through the "quick fix." The process that could address the fundamental source of the problem–lack of a compliance mentality–would involve collaboration with the agency. Unfortunately, due to all the challenges the company has put the agency through, the agency is more inclined to be hostile than cooperative. Moreover, even if the agency were cooperative, it would take time for the enhanced compliance mentality to manifest itself in reduced compliance deficiencies. This delay in feedback between the action and the result means that people are more likely to pay attention to results that are more immediately linked to the action–in this case, the quick fix.

The triumph of the short-term quick fix over the long-term fundamental fix is well known to environmental professionals. It explains why many well-meaning companies can't find their way out of a "fire-fighting" pattern, even when expressing deep frustration with their inability to behave otherwise. It explains why many efforts that raise the environmental awareness of line management fail to take root, and why many business managers view investments in cost avoidance with skepticism.

The good news is that a desire to improve the situation coupled with an understanding of the systems dynamics can lead a company to break out of a negative loop. The company in this example made substantial progress in reducing its compliance deficiencies after becoming aware of its knee-jerk dependence on the quick fix. By issuing instructions to its legal staff to take a more collaborative approach to dealing with regulatory agencies, it helped erode the reinforcing power of the quick-fix loop. It created financial incentives for its managers to reduce compliance deficiencies and to implement new programs designed to create future compliance. The company also hired more environmental staff whose mission was to raise the awareness of line management and put in place management systems that would head off future compliance issues. And it took steps to repair the damaged relationship with the regulatory agency by enlisting one of its manufacturing plants as the pilot site for a pet project of the agency. All of these steps, coupled with a well communicated and understood analysis of the systems dynamics, led the company out of what seemed to be a hopeless quagmire.

Once an implementation plan is developed, with due attention paid to measurement and organizational considerations, the next step is to make it happen. Strategy and plans cannot create business value without implementation.

VI.

Implementation: Lessons Learned in the Field

THE DISCUSSION IN CHAPTER 5 focused primarily on measurement and organizational considerations in developing a plan to implement the strategy. But, like any important project, implementation requires definition of tasks, assignment of roles and responsibilities, setting of schedules, establishment of milestones, and allocation of resources. Perhaps the most important aspect of implementation is the need to sustain momentum. This chapter presents a case study of how Monsanto Company is sustaining the momentum of its bold new environmental business strategy.[17] Monsanto's approach provides some helpful lessons, including the need to develop a ground swell at the grassroots, to concentrate on critical tasks, to bring tools up to speed quickly and to continue to expand and challenge current thinking.

A Snapshot of Monsanto's Progress

In 1996, Monsanto was a science-based company, 95 years old, with over 30,000 employees devoted to the discovery, manufacturing and marketing of agricultural products, performance chemicals used in consumer products, prescription pharmaceuticals, and food ingredients. In 1997, the company split into two companies: one focused on life sciences and the other on chemicals. The life sciences company, Monsanto, comprises agricultural, pharmaceutical and nutrition businesses, while the chemicals company, named Solutia, focuses on traditional businesses in fibers and specialty chemicals. The new Monsanto, the life sciences company, plans to emphasize sustainable development in everything it does.

In the early 1990s, Monsanto's then CEO, Richard Mahoney, began actively pursuing environmental goals when he pledged to reduce hazardous air emissions by 90% against the 1987 base emissions level. The goal came as both a shock and a wake-up call to company employees and other stakeholders. Yet, through the growing pains of reaching for environmental goals, the company realized there are revenue opportunities in products and technologies that are good for the environment.

Monsanto's management recognized that global issues emerging around feeding a growing population, losing topsoil and increased pressure on the planet's finite fresh water resources might provide new opportunities. Senior management began to ask questions like this one: How could Monsanto become more efficient, improve crop yields, decrease company inputs and make a net gain?

Monsanto has now made sustainable development one of the key elements in its new statement of corporate values. One of its five restructured business groups is called Sustainable Development. That group's objectives are twofold: to provide strategic context and tools to facilitate Monsanto's move toward greater sustainability, and to develop profitable new businesses that are consistent with the principles of sustainable development. According to Kate Fish, Monsanto's director of sustainable development, the group's core capabilities are:

- Integration of knowledge, insights, framework and measurement systems across the company in order to position all the businesses to thrive in pursuit of a sustainable future
- Application of our understanding of emerging discontinuities to create new business opportunities and generate enormous competitive advantage
- Enhancement of our freedom to operate through alignment of business objectives with meeting world needs

Monsanto's Methods for Implementation

To understand how Monsanto is moving forward to implement its new strategy, we examine here several aspects of the company's approach: design, roles and responsibilities, resource deployment, communication, partnerships, monitoring progress, and tools and pilot programs.

Design

The design of Monsanto's strategy began with a cross-functional brainstorming session. Monsanto's chief executive officer, Robert Shapiro, asked a group of about 25 critical thinkers–including management board members; business unit leaders; and people from planning, manufacturing, and safety and health–to "go off, think about what's happening in the world, and come back with some recommendations about what it means for Monsanto." Representatives from outside the company were also invited to participate and challenge Monsanto's thinking and assumptions about the world.

That off-site meeting fired up those involved to work hard to contribute to a more sustainable future. The company began to explore the following areas:

- **Corporate impact.** What was Monsanto's comprehensive impact on the environment? How could Monsanto understand its total throughput, in addition to toxic releases?
- **Indicators.** How could Monsanto rate the impact of its services and products on the environment? What are the right metrics?
- **Eco-efficiency.** How does Monsanto utilize its resources? How efficient are its processes?
- **Opportunities.** What new businesses and services will help Monsanto meet people's needs in less resource-intensive ways?

CEO Bob Shapiro articulated to the employees that sustainable development was important to help drive the company toward the future. The initiative formally began as a volunteer activity, and business units were solicited for input or participation. A wave of e-mail went around the world and about 70 staff volunteered. An off-site meeting was held to develop an implementation strategy and to organize the company to add value and generate business through sustainable development. This meeting led to the establishment of Monsanto's seven sustainability teams focused around:

- **Methodologies and tools:** sustainability indicators, eco-efficiency and full-cost accounting
- **New business opportunities:** new businesses/new products, water and global hunger
- **Education:** communication and education

Through the teams' efforts, the company's business strategies are beginning to align with the needs of sustainable development. Monsanto strategic business units have started to incorporate sustainability into their long-range plans and capital-planning processes.

Roles and Responsibilities

The teams were initially organized as a volunteer effort. A core group signed on, but as word spread, more members than expected stepped forward. Even as some staff got involved in other engagements and left the team, typically four or five active and committed members remained.

The teams were not organized with a script. According to CEO Bob Shapiro, "You have some ideas, some activities, some exhortation, and some invitations, and you try to align what people believe and what people care about with what they're free to do." The teams were neither a top-down nor a bottom-up exercise. Monsanto encouraged creativity to "see what ideas really win people's hearts and trust that those ideas will turn out to be the most productive."

Resource Deployment

For an initiative that was in its early levels of exploration, Monsanto chose to rely for more that a year on a subjective sense of momentum. During that initial phase, the teams prepared budgets for activities, consultants, and tools. Because of the overarching support from the CEO and the strategic importance of the initiative, the teams' budgets were generally accepted. Today the company is setting quantitative goals, macro budgets and timelines.

Communication

Communication throughout the company was generated through the intense momentum and excitement of the employees. CEO Bob Shapiro continues to articulate his vision and commitment. Business unit managers are solicited for their input and participation. The company Web site, updated often, identifies sustainable development as a core driver of value for the company. The Annual Environmental Report extensively details the company's progress and is easily downloaded from Monsanto's Web site. To educate employees and stimulate the pursuit of initiatives that link profits and environmental benefits, the Communication and Education team provides a four-hour training course with case studies and breakout sessions. The training also challenges employees to think more broadly about their business areas. And, externally, response from the new media has been extremely encouraging, including supportive profiles in business magazines, trade journals and the Harvard Business Review.

Partnerships

A key competitive advantage of Monsanto's approach is the company's openness to learn from others, including the desire to partner with nongovernmental organizations and small businesses, thus tapping into the perspectives of a wide range of people. For example, the company organized a roundtable on Sustainable Business Opportunities with 16 experts from 13 countries, with the intention to publish the outcome in Monsanto's Environmental Annual Review. The event was unusual because it had potential for the expression of dissent about Monsanto products and policies. In fact, collaboration with groups that have in the past criticized Monsanto has caused some worry, but Monsanto also sees the activity as a vehicle to learn from and contribute to the larger debate and policies surrounding sustainable development.

Monitoring Progress

Monsanto's intent has been to rely on the energy and interests of its employees and to find dedicated people in each business area to help develop and apply tools and metrics. In addition, a number of people in the Sustainable Development Sector are dedicated to and have their performance evaluated based on the company's progress. In fact, a key factor in the design of the initiative is to fully engage the employees: "Companies aren't machines anymore. We have thousands of independent agents trying to self-coordinate because it is in their interest to do so," stated CEO Bob Shapiro.

Tools and Pilot Programs

Three Monsanto teams were designed to develop tools that will cut across the company and become a framework for decisions related to sustainable development. Monsanto has engaged external consultants, including Arthur D. Little, to assist in designing these tools. The sustainability screening matrix, eco-efficiency and full-cost accounting model are each being tested in pilot projects before being rolled out across the entire company.

Lessons Learned

Though analyzing current and future business opportunities is not an activity unique to Monsanto, the company's environmental strategy is truly bold and different. We can identify ten lessons, in particular, from which companies interested in pursuing similar but different pathways can learn.

1. **Implement your environmental strategy as a business initiative.** Monsanto is implementing its sustainable development initiative as a framework and driver for new business opportunities and growth.
2. **Have a firm commitment from the chief executive officer or board chair.** Monsanto's vision has been championed and articulated by CEO Bob Shapiro. His words and leadership send a clear message to the employees about the commitment to and strategic importance of pursuing sustainability.
3. **Develop a groundswell at the grassroots.** Though new within Monsanto's culture, this grassroots approach to engaging employees' interests and creativity has helped Monsanto percolate with ideas.
4. **Develop a way to convince middle management.** This has been one of Monsanto's steepest challenges. Because middle managers have tough production and sales goals, they typically look for short-term, value-added results. To approach this group, it is important to look for meaningful short-term objectives that can be widely and quickly disseminated throughout the organization. Also consider building environmental performance into the performance measurement of the managers.

5. **Get the perspectives of top thinkers.** Developing an environmental strategy is a value-added process. Key input from experienced people can go very far in promoting thinking and ensuring progress. If Monsanto were given 10% more resources for developing its sustainable development thrust, at least some of the money would go toward seeking collaborations and partnerships with experts and leading organizations outside the company, according to Kate Fish.

6. **Concentrate on the critical tasks and bring your tools up to speed quickly.** The most critical of Monsanto's seven sustainability teams were those focused on developing sustainability and eco-efficiency matrices. Any delay in developing these important tools affects timelines and progress. When necessary, consider using outside resources to jumpstart the development of critical tools.

7. **Expand and challenge your thinking.** Key to Monsanto's approach is the recognition that the company does not have all the answers. Though obviously Monsanto does not intend to implement every suggestion made, the company continues to look for full understanding of the challenges at hand and to engage in critical dialogue with others.

8. **Ensure that there are committed, motivated members on each team.** A core of dedicated members, authorized to own the issues at hand, is critical to ongoing momentum and team success. At Monsanto, team members' performance evaluation is based on the team's progress. And, the head of one team is also the coordinator of the activities of other teams, which ensures collaboration.

9. **Educate your employees.** The link between the value of environmental benefits and company profits is not necessarily intuitive to employees. For example, employees need to step back and understand how an environmental lens can enhance sales and production. Monsanto provides employee education through a four-hour training course about the business benefits of sustainability.

10. **Design practical programs mindful of employees' needs.** Monsanto's original education module around sustainability stumbled because it was two days long–more time than employees wanted to spend. The revised four-hour course successfully engages employees in the core concepts without taking too much of their time.

Afterword: How to Begin the Process

For a company that already has in place a vision for achieving environmental performance excellence, and programs to support that vision, the steps toward creating business value are well within reach. In fact, I would be very surprised if any company achieving increasing levels of environmental performance is not already contributing to increased business value–even if the company is not capturing that value in business terms.

The intent of this book has been to illustrate that there is a stepwise process toward blending together the two shades of green: environmental value and business value. And though the steps may vary somewhat in the context of a particular company, its culture, its strategic vision, and the issues driving its industry, there are common important themes that should be embraced, whatever the context. Let me review briefly those I consider the most important (and refer you back to where they are discussed in the book).

- Ask the question: "What is important to our customers, shareholders, employees and the communities in which we operate?" The answer is a key factor in defining what "business value" means for the company. (Pages 1–2, Chapter 1)

- Develop an expansive, robust understanding of the full set of environmental costs borne by the company because that's the foundation of the case that good environmental management provides business value. (Pages 6–8, Chapter 1)
- Remember, accounting for environmental costs is only half the battle. To affect the bottom line, use this information to make better decisions for product design, process design and management, facility siting, purchasing, operations, risk management, capital investments, cost control, waste management, cost allocation, product pricing, product retention and mix, and profitability. (Page 9, Chapter 1)
- The strategy process begins with vision, an image of a desired future. Ask the question, "What do we want to create?" Make sure that the answer is both aspirational and inspirational. (Page 13, Chapter 2)
- The vision should be shared by, not with, individuals in the organization. Arrive at an environmental vision in a collaborative, inclusive, bottom-up way. (Page 15, Chapter 2)
- Assess the company's current situation in light of its long-term vision and appreciate the gap between the two. An essential element of this gap is the tension that arises in understanding how far the organization is from achieving its vision. This tension is a positive and creative force that draws everyone toward the vision. (Page 19, Chapter 3)

- The extent to which individual and group goals align with the overall corporate vision determines whether the organization succeeds or fails. It is critical to assess the extent to which environmental objectives are aligned with business objectives (Page 22, Chapter 3)
- Use the value chain systematically to analyze each value-adding business activity, from design to end of life, and to find ways for environmental management to contribute to business objectives. Then, look for a healthy balance of focus in the value portfolio. (Pages 23–24, Chapter 3)
- Find ways for environmental management to contribute to the company's ability to meet strategic objectives for both growth and innovation–and you will hit a business-value home run. (Page 26, Chapter 3)
- Identify and assess the environmental issues that might impact the company's competitive position, both near and long-term. Develop scenarios based on the issues to consider alternative plausible futures that go beyond extrapolating from the present. (Pages 36–41, Chapter 4)
- Make sure that the environmental strategy enables the company to do different things than its competitors, and to do things differently than competitors. Visit the concept of "Industrial Darwinism" to help think about strategic choices. (Pages 41–45, Chapter 4)

- Don't underestimate the potential value of operational effectiveness in environmental processes. Form "tighter loops": move away from thinking linearly to seeing interrelationships and the way things work in a systemic manner. (Pages 51–54, Chapter 4)
- Experience shows that effective processes are based upon work that is performed across functional boundaries. Align the right set of diverse talents and perspectives to achieve a common purpose. (Page 57, Chapter 5)
- Look to Design for Environment approaches to learn a range of lessons in implementation planning. (Page 58, Chapter 5)
- Develop the right metrics to show how environmental activities contribute to the company's business objectives–this cannot be stressed too strongly. (Pages 60–61, Chapter 5)
- Tearing down the "green wall" between environmental and business managers requires the efforts of both parties. Environmental managers need to speak the business language and business managers need to become more knowledgeable and accountable. (Pages 64–67, Chapter 5)
- Use the discipline of systems thinking to understand and break out of destructive patterns of behavior. (Pages 67–70, Chapter 5)

- Make sure implementation is sustained. Learn from Monsanto. Though the company's environmental strategy is unique, the lessons in how the company planned and implemented its strategy are valuable for all of us. (Pages 71–81, Chapter 6)

APPENDIX

This Appendix contains a tool for evaluating and assessing the strengths and weaknesses of a company's environmental, health, and safety (EHS) management processes. It represents the learning and experience gained by Arthur D. Little professionals in over 20 years of assessing the EHS management processes and systems of companies around the world. The tool consists of 21 sets of evaluative criteria that can be applied to assess the 19 EHS management processes depicted in the figure below and in the book's Figure 6 (see page 28), plus alignment criteria around organization and resources. The criteria used are *process* oriented. They focus on such things as efficiency, speed/cycle time, degree of integration with the business, consistency, continuous improvement, learning and flexibility. Use of the rating scale to arrive at a score is optional.

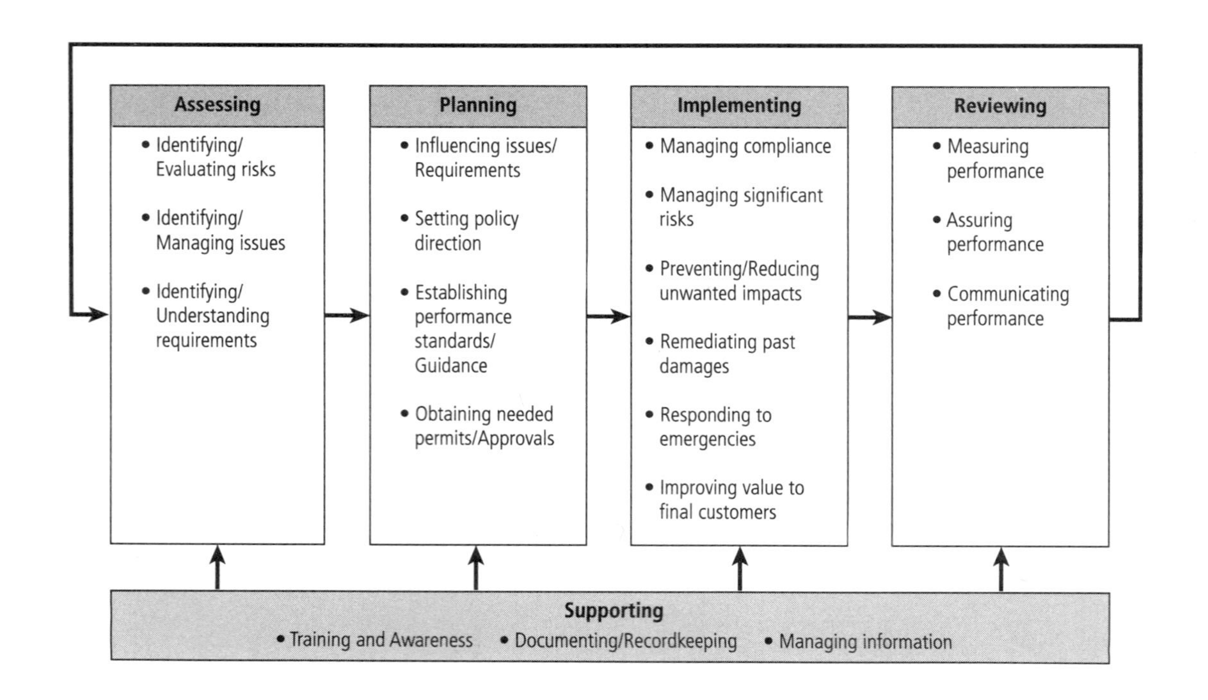
Assessing
• Identifying/Evaluating risks
• Identifying/Managing issues
• Identifying/Understanding requirements
Planning
• Influencing issues/Requirements
• Setting policy direction
• Establishing performance standards/Guidance
• Obtaining needed permits/Approvals
Implementing
• Managing compliance
• Managing significant risks
• Preventing/Reducing unwanted impacts
• Remediating past damages
• Responding to emergencies
• Improving value to final customers
Reviewing
• Measuring performance
• Assuring performance
• Communicating performance
Supporting
• Training and Awareness • Documenting/Recordkeeping • Managing information

Identifying/Evaluating Risks

Management Process Assessment and Rating

	①	②	③	④	⑤
Process Attribute	Needs Improvement		Meets Threshold Expectations		Significant Strength
Clear Ownership, Roles and Responsibilities	EHS risk assessment roles and responsibilities are not clearly defined.		EHS risk assessment is largely staff driven. Risk assessors have limited background and experience in assessment techniques.		Line organization drives EHS risk-assessment activity. Assessments are performed by highly qualified personnel.
Appropriate Scope/Coverage	Process focuses on routine risks rather than the full range of possible operating conditions, on operations rather than the full product life cycle, and on internal risks only, not third parties.		Process covers the major EHS risks facing the business; ensures that new or changing risks are identified and appropriately evaluated.		Process addresses the full range of relevant EHS risks, including emerging issues. Business management has a detailed understanding of the EHS risks it faces.
Speed/Cycle Time	Information on significant EHS risks typically not available in a timely fashion for business decision-making.		EHS risk information is generally timely for business decision-making.		Process is streamlined, routinely allows business managers to consider EHS risks as part of key decisions.
Consistency/ Reliability	Methodologies used are informal, results are frequently undocumented and difficult to replicate.		Methodologies are generally appropriate to the EHS risks being considered.Assess-ments are conducted by qualified professionals.		Rigorous, well-documented assessments of EHS risks yield highly reliable replicable results that are well-accepted by business managers.
Integration with Other Key Processes	Risk identification and evaluation activities isolated from key business planning and decision-making processes.		Major business issues such as new projects and acquisitions/divestitures receive appropriate EHS risk evaluation.		EHS risk identification and evaluation is an integral part of business planning and decision-making, e.g., strategic and operations planning, R&D, new product development, supply chain management.
					Overall Rating:

Identifying/Managing Issues

Management Process Assessment and Rating

	① ②	③	④ ⑤
Process Attribute	Needs Improvement	Meets Threshold Expectations	Significant Strength
Clear Ownership, Roles and Responsibilities	Roles and responsibilities for tracking emerging issues are nonexistent or not clearly defined.	Roles and responsibilities are defined and well understood for tracking emerging EHS issues but less so, if at all, for managing EHS issues. Issue management is reactive at best.	Roles and responsibilities for both tracking and managing emerging EHS issues are defined and well understood. EHS issues are actively managed.
Appropriate Scope/Coverage	Issues addressed on a hit-or-miss basis, if at all.	Issue identification and management tends to focus on manufacturing/ operating issues only.	Issues are identified and managed effectively across the entire product life cycle.
Involvement of the Line Organization	Senior and line operating management unaware of potentially significant EHS issues affecting the business.	Senior and line operating management show awareness of potentially significant EHS issues. Limited involvement, if any, in issues management.	Senior and line operating managers drive the management of potentially significant EHS issues affecting the business.
Integration with Other Key Processes	Little or no integration of EHS issues management into other key business processes.	Some evidence of integration of EHS issues management with governmental/external affairs.	Evidence of integration of EHS issues management extends to strategy/planning, R&D/new product development, marketing and others.
			Overall Rating:

Identifying/Understanding Requirements

Management Process Assessment and Rating

	①	②	③	④ ⑤
Process Attribute	Needs Improvement		Meets Threshold Expectations	Significant Strength
Clear Ownership, Roles and Responsibilities	Roles and responsibilities are nonexistent or not clearly defined.		Roles and responsibilities are generally well-defined, but overlaps and conflicts exist among multiple staff units.	Roles and responsibilities are clearly defined; any overlaps or conflicts are minimized.
Consistancy/ Reliability	Some requirements are missed or identified only after the fact.		Legal/regulatory requirements are identified consistently. Some inconsistencies in application across the organization.	Legal/regulatory as well as voluntary requirements are consistently identified and applied.
Speed/Cycle Time	Typically, little time, if any, given to influence developments and/or prepare implementation plans.		Interpretations generally timely for implementation planning; not so for influencing developments.	Interpretations are sufficiently early to allow effective influence on developments, when appropriate.
Efficiency/ Cost-Effectiveness	Every organization level and individual business unit tends to replicate each other's efforts.		Process works generally well, but still involves considerable overlap and duplication of staff effort.	Process is streamlined, minimum number of staff involved, sufficient to the purpose and effectively leveraged.
				Overall Rating:

Influencing Issues/Requirements

Management Process Assessment and Rating

	① ② ③ ④ ⑤		
Process Attribute	Needs Improvement	Meets Threshold Expectations	Significant Strength
Clear Ownership, Roles and Responsibilities	Industry/trade association participation ad hoc, based on individuals' interests. No clear definition of roles/ responsibilities for independent influence efforts.	Industry/trade association involvement is managed effectively. Overlapping roles/responsibilities among multiple units for independent efforts.	Industry/trade association involvement and independent influence efforts are well managed, to the organization's clear benefit.
Integration with Other Key Processes	Influence efforts largely an EHS functional concern. Little or no visibility elsewhere in the organization.	Multiple functional are involved units (government/external affairs, legal, others), but leadership and coordination still suffer.	Managed as an integrated process focused on the organization's overall interests.
			Overall Rating:

Setting Policy Direction

Management Process Assessment and Rating

	①	② ③	④ ⑤
Process Attribute	Needs Improvement	Meets Threshold Expectations	Significant Strength
Clear Ownership, Roles and Responsibilities	No involvement by senior executives or operating management in development of the policy. No clear alignment with business strategy or management's views.	Some involvement by executive and operating management in policy development. Policy statement not inconsistent with management's basic views.	Strong senior leadership in helping frame the basic policy. Closely aligned with organization's overall business strategy and vision.
Appropriate Scope/Coverage	Policy statement fails to address significant areas of the organization's EHS potential risks and exposures.	Policy statement addresses most areas of potentially significant EHS risk and exposure at a generally appropriate level for the organization.	Policy statement provides a clearly articulated posture regarding all potentially significant EHS risks and exposures facing the organization.
Continuous Improvement/ Learning	Existing policy statement outdated, not reviewed to reflect changing needs and risks of the business.	Policy statement reasonably current. No process for reviewing the policy on a regular basis.	Policy statement is a living document that reflects current business needs and risks and is reviewed and revised periodically.
			Overall Rating:

Establishing Performance Standards/Guidance

Management Process Assessment and Rating

	1	2	3	4	5
Process Attribute	Needs Improvement		Meets Threshold Expectations		Significant Strength
Clear Ownership, Roles and Responsibilities	No corporate first-level EHS guidance exists. Facilities generally do not manage against formal EHS goals, plans, objectives or targets. Facility EHS procedures are weak or nonexistent.		Some corporate first-level EHS guidance in place. Limited facility-level EHS goals and plans exist, but accountability is weak. Facility EHS procedures exist but are not integrated into SOPs.		Strong corporate first level EHS guidance exists, well accepted by facilities. Effective facility EHS management stresses accountability against measurable objectives and targets, integration of EHS into operations management.
Involvement of the Line Organization	Little or no input from division and facility operating management in development of EHS policy guidance, facility-level EHS goals and plans.		Some limited input from division and facility operating management in EHS policy guidance, facility plans.		Strong input and acceptance from division and facility operating management regarding EHS policy guidance, facility plans.
Clear/Accurate Performance Measurement	Neither first-level EHS guidance nor facility-level management style address EHS performance measurement.		First-level EHS guidance calls for performance measurement, but facility-level follow-through is limited.		First-level EHS guidance stresses performance measurement, reflected in strong facility-level measurement and accountability systems.
					Overall Rating:

Obtaining Needed Permits/Approval

Management Process Assessment and Rating

	①	②	③	④	⑤
Process Attribute	Needs Improvement		Meets Threshold Expectations		Significant Strength
Clear Ownership, Roles and Responsibilities	EHS permitting roles and responsibilities are not clearly defined.		EHS permitting is largely the preserve of the EHS staff. Roles and responsibilities of other affected functions are unclear.		EHS permitting roles and responsibilities are clearly defined, including all other affected functions.
Involovement of the Line Organization	Little or no involvement by affected line managers in EHS permit development/negotiation.		Affected line managers' views are frequently solicited, but the permitting process is still largely driven by EHS technical, functional considerations.		Affected line managers' views are integrated into permitting strategy and permit development, and they help drive the overall process.
Integration with Other Key Processes	Little or no coordination of EHS permitting activities with key related processes; e.g., planning, capital budgeting, project management, engineering, legal.		Some coordination with key related processes but largely ad hoc based on EHS staff's personal relationships and networks.		Strong systematic coordination with other key related processes, providing them with appropriate influence.
Speed/Cycle Time	Little or no lead time in identifying permit needs. Permit development and negotiation process is slow, frequently resulting in significant costs to the organization.		Lead times are acceptable for identifying permit needs. Permit development and negotiation process is still slow and costly to the organization.		EHS permit needs are identified early. Permit development and negotiation process has been streamlined to reduce delay, cost.
					Overall Rating:

Managing Compliance

Management Process Assessment and Rating

	①	②	③	④ ⑤
Process Attribute	Needs Improvement	Meets Threshold Expectations	Significant Strength
Involvement of the Line Organization	EHS is staff driven; little sense of line accountability for results.	EHS is staff led, but some line ownership of results exists.	Line-driven, line ownership and accountability for results.
Consistency/ Reliability	Uneven, unreliable compliance.	Mainly in compliance; exceptions are persistent but minor.	Compliance is routinely achieved; strong systems are in place.
Flexibility/Handling of Exceptions	Difficult to distinguish potentially significant compliance issues.	Potentially significant compliance issues are highlighted.	Potentially significant compliance issues receive appropriately expedited handling, resolution.
Cost Efficiency	Heavy EHS staff use, process is not well defined; every time is nonroutine.	Basic process defined; still involves considerable drain on EHS staff time.	Process streamlined; operations staff is well-leveraged.
Continuous Improvement/ Learning	Persistent repeat findings; little evidence of improved compliance.	Fewer repeat findings over time.	Corrective actions address root causes; little evidence of repeat findings.
			Overall Rating:

Managing Significant Risks

Management Process Assessment and Rating

	①	②	③	④	⑤
Process Attribute	Needs Improvement		Meets Threshold Expectations		Significant Strength
Involvement of the Line Organization	Line organization has low awareness of EHS risks of its operations. Line personnel view EHS risk management as someone else's job.		EHS risks are well understood by line organization. Line personnel accept management responsibility.		EHS risk management is routinely integrated into standard operating procedures. Strong line sense of EHS risk management responsibility.
Speed/Cycle Time	EHS risk reviews are frequently performed too late to influence business decisions.		EHS risk reviews are generally timely to the business decision purpose.		EHS risk reviews are routinely performed early in the business decision-making process.
Integration with Other Key Processes	EHS risk reviews are done mainly for major capital projects only.		EHS risk reviews focus primarily on operations issues.		EHS risk reviews address the full spectrum of business issues across the product life cycle.
Continuous Improvement/ Learning	Information on actions taken to manage and/or minimize EHS risks is not widely shared within the organization.		Some efforts are made to share EHS risk management approaches within the organization.		Formal system in place to ensure that EHS risk management activities are communicated within the organization and implemented wherever appropriate.
					Overall Rating:

Preventing/Reducing Unwanted Impacts

Management Process Assessment and Rating

	①	②	③	④	⑤
Process Attribute	Needs Improvement		Meets Threshold Expectations		Significant Strength
Clear Ownership, Roles and Responsibilities	Impact prevention/ reduction is not identified as an activity to be managed. Roles and responsibilities are not clearly defined and assigned.		Impact prevention/ reduction is recognized as an important activity. Specific prevention/ reduction initiatives are driven largely by EHS staff.		Impact prevention/ reduction is viewed as a key line operating responsibility. Incentives exist and are effective in promoting line management initiatives in this area.
Integration with Other Key Processes	Prevention/reduction activities are either nonexistent or conducted largely separately from business operations and decision making.		Prevention/reduction activities linked to business decisions on new or modified manufacturing processes.		Impact prevention/ reduction efforts are integrated into the full range of business planning and decision-making processes, across the product life cycle.
Continuous Improvement/ Learning	Impact prevention/ reduction activities are not viewed as a quality issue; no emphasis on continuous improvement.		Specific targets are established for impact prevention/reduction over time, and initiatives are managed against them.		Impact prevention/ reduction efforts are tied into and recognized as an integral part of the organization's quality management system.
Utility, Value of Results	Few, if any, examples of successful EHS impact prevention/reduction exist.		Some demonstrable local successes in achieving measurable impact reduction and realizing cost savings.		Strong overall record of achieving significant impact reductions and cost savings throughout the organization.
					Overall Rating:

Remediating Past Damages

Management Process Assessment and Rating

	①	②	③	④	⑤
Process Attribute	Needs Improvement		Meets Threshold Expectations		Significant Strength
Clear Ownership, Roles and Responsibilities	Remediation is a collateral duty of professionals having other primary responsibilities. Remediation process is not well defined.		Remediation activity is managed by dedicated professionals, subject to a well-defined process that identifies clear roles and responsibilities.		Well-resourced and well-managed remediation activity receives strong oversight and policy direction from senior business management.
Cost Efficiency	Remediation costs are not tracked, projected, or periodically reviewed. Outsourcing opportunities have not been fully explored.		Remediation costs are well-managed based on site-specific cleanup strategies that pursue cost-effective remedial standards.		Remediation process is streamlined to eliminate low value-added activities. Contractors are given incentive to innovate, improve effectiveness and control costs.
Continuous Improvement/ Learning	Little or no evidence that the remediation activity gets smarter, faster or more efficient with more sites under its belt.		Remedial site experience is well documented and reviewed for improvement opportunities. Performance measures show improvement over time.		A well-defined process promotes active exchange of learning across individual remedial sites. Contractors' experience is tapped effectively.
					Overall Rating:

Responding to Emergencies

Management Process Assessment and Rating

	① ② ③ ④ ⑤		
Process Attribute	Needs Improvement	Meets Threshold Expectations	Significant Strength
Clear Ownership, Roles and Responsibilities	Site emergency response roles and responsibilities are not clearly defined or well understood.	A site emergency response team is designated. Team members have clearly defined, well-understood roles and responsibilities.	Site emergency response team members and suitable back-ups are well trained and drilled. Effective coordination exists with local authorities and off-site response/service providers.
Appropriate Scope/Coverage	Site emergency response plan is generic or at best cursory, out-of-date and otherwise of little use.	Site emergency response plan addresses principal identified site hazards and reasonably foreseeable emergency scenarios.	Site plan also anticipates less likely but still plausible emergency scenarios and is consistent with existing local and area-wide emergency response plans.
Continuous Improvement/ Learning	No regular tests or drills, plan document is rarely updated.	Site emergency response plan procedures are regularly tested and drilled. Plan document is updated to reflect site changes and lessons learned.	Site emergency response plan procedures are frequently tested and drilled. Plan document is evergreen, including incorporating lessons learned from other sites across the organization.
			Overall Rating:

Improving Value to Final Customers

Management Process Assessment and Rating

	①	②	③	④	⑤
Process Attribute	Needs Improvement		Meets Threshold Expectations		Significant Strength
Clear Ownership, Roles and Responsibilities	EHS roles and responsibilities for new product development and product stewardship are not clearly defined.		EHS considerations are represented in new product design process. Product stewardship activity exists but is largely EHS staff driven.		Effective integration of EHS considerations into new product development process. Clear sense of line ownership of product stewardship process, vigorously implemented.
Integration with Other Key Processes	EHS management systems are separate from those for new product development, product safety.		EHS management systems are clearly linked to those for new product development and product stewardship, and they deliver identifiable business benefits.		EHS management systems are fully integrated with effective business processes for new product development and product stewardship.
Customer Value	Little demonstrated market interest or value attached to organization's current efforts to sell EHS-advantaged products, communicate product risks.		Some limited demonstrated market interest and value attached to certain specific efforts to sell EHS-advantaged products or communicate product risks.		Considerable business benefits accruing to aggressive EHS-advantaged new product development and product stewardship activities.
					Overall Rating:

Measuring Performance

Management Process Assessment and Rating

	① ② ③ ④ ⑤		
Process Attribute	Needs Improvement	Meets Threshold Expectations	Significant Strength
Appropriate Scope/Coverage	Few, if any, EHS performance measures exist. Developed only where required by government regulation.	EHS performance measures address regulatory standards and key company goals and objectives.	Performance measures are developed to focus on full range of EHS issues/risks, enabling managers to get comprehensive picture of EHS performance.
Involvement of Line Organization	Little involvement of line organization in designing, collecting or using EHS performance information.	Line organization plays an active role in the various aspects of the performance measurement process.	Line organization drives the performance measurement process, playing the critical role in every aspect of performance measurement.
Consistency/ Reliability	Few, if any, procedures to ensure consistency, accuracy and reliability of EHS performance data. Quality of performance information varies by measure and source of data.	Systems/procedures in place to produce consistent, reliable EHS performance information in most cases.	Process consistently produces performance information that is recognized and valued by managers for its accuracy and reliability.
Integration with Other Key Processes	Little, if any, integration of performance measurement with other key EHS and business management processes.	Some integration, mainly with related EHS management processes and operations management.	EHS performance measurement fully integrated with EHS and business management processes across the product life cycle and throughout related overhead activities.
			Overall Rating:

Assuring Performance

Management Process Assessment and Rating

	①	② ③	④ ⑤
Process Attribute	**Needs Improvement**	**Meets Threshold Expectations**	**Significant Strength**
Clear Ownership, Roles and Responsibilities	No formal independent EHS audit activity exists.	Independent EHS audit activity exists but is largely staff driven. Line accountability for correcting identified problems is weak.	Line operating managers demand rigorous, independent EHS audits; regarding themselves as fully accountable for correcting identified problems.
Appropriate Scope/Coverage	EHS audits address legal compliance requirements only.	EHS audits also address company policies/ procedures. EHS management systems not routinely assessed, if at all.	Rigorous EHS compliance and management systems audits are conducted periodically.
Consistency/ Reliability	No process exists to ensure that audit results are consistent and reliable.	The company uses an effective combination of procedures, training and skilled auditors to promote audit results that are generally consistent and reliable.	The company periodically verifies the consistency and reliability of audit results. Line managers feel confident in relying on audit results.
Speed/Cycle Time	Audit results are often not communicated until long after the audit is performed. No effective process exists for ensuring that deficiencies are corrected in a timely manner.	Audit results are typically communicated within a reasonable time after the audit is completed. Deadlines are developed and tracked for all corrective actions.	Cycle time for audit results and corrective actions is measured and managed to encourage timely performance and continuous improvement.
Continuous Improvement/ Learning	Audits are not used to drive continuous improvement. Little or no analysis of root causes or trends in audit findings; pattern of frequent repeat findings.	Audit results are used to help drive continuous improvement. Some limited use of root cause analysis but no process exists for sharing audit learnings more widely.	Line operating managers use audit results to actively promote continuous improvement efforts. Root cause and periodic trend analysis performed routinely. Process in place to share audit learnings widely.
			Overall Rating:

Communicating Performance

Management Process Assessment and Rating

	①	②	③	④ ⑤
Process Attribute	Needs Improvement		Meets Threshold Expectations	Significant Strength
Clear Ownership, Roles and Responsibilities	Roles and responsibilities for internal/external EHS communications not clearly defined.		EHS-related status and performance information routinely reported to line operating managers.	EHS-related status and performance information routinely reported across functions and among organizational levels. Proactive communication of this information externally.
Consistency/ Reliability	EHS-related information collected and reported ad hoc, differently in different parts of the organization.		Some standardization of key EHS-related information, routinely reported. Not rolled up to corporate level.	EHS-related information fully standardized and consistently reported at every organizational level, including corporate.
Degree of Automation	Little or no automation of EHS-related data collection, analysis and reporting.		Some use of automated systems to collect, analyze and report EHS-related information, but full potential for use still not explored.	Innovative and effective use of information technology approaches to enhance communication of EHS performance.
Integration with Other Key Processes	No formal EHS-related databases.		EHS-related databases are wholly separate from those for operations (e.g., production levels, raw material use).	EHS-related databases and operations databases are effectively integrated or otherwise linked.
Continuous Improvement	EHS-related information and reporting does not address its use for promoting continuous improvement.		EHS-related information is used to drive continuous improvement for selected activities in certain parts of the organization.	EHS-related information is used to drive contin-uous improvement uniformly across the organization.
				Overall Rating:

Training and Awareness

Management Process Assessment and Rating

	①	② ③ ④	⑤
Process Attribute	Needs Improvement	Meets Threshold Expectations	Significant Strength
Clear Ownership, Roles and Responsibilities	Roles and responsibilities for EHS training are not clearly defined or well understood.	EHS staff largely drives the EHS training activity. Line management accepts little responsibility.	Line management drives the EHS training activity and accepts full responsibility and accountability for its effectiveness.
Appropriate Scope/Coverage	No formal EHS training needs assessment has been conducted. Training done is hit-or-miss.	EHS training is largely focused on meeting legal and regulatory requirements. Awareness activities are minimal and ineffective.	EHS training addresses the full range of EHS needs and issues. Awareness activities are strong and frequently reinforced.
Integration with Other Key Processes	Little or no integration of EHS training activity with other key EHS management processes.	EHS training activity helps to reinforce compliance-related processes, i.e., identifying/understanding requirements, managing compliance.	EHS training activity is also integrated closely with EHS processes for identifying management issues, managing risks and documenting/record-keeping.
Continuous Improvement/ Learning	EHS training does not address lessons learned from operating experience. No emphasis on performance improvement.	EHS training may address improving EHS compliance performance, audit scores.	EHS training follows the quality model, identifying and capitalizing on opportunities for improving EHS performance across the board.
			Overall Rating:

Documenting/Recordkeeping

Management Process Assessment and Rating

	① ②	③	④ ⑤
Process Attribute	Needs Improvement	Meets Threshold Expectations	Significant Strength
Clear Ownership, Roles and Responsibilities	No functioning systems in place for EHS document control and record-keeping. Roles and responsibilities are undefined.	Generally effective systems exist for EHS document control and record-keeping, but they may not be formal or documented.	Effective, well-documented systems drive clear roles and responsibilities for EHS document control and record-keeping.
Simplicity	EHS record-keeping is left to individual initiative and discretion, making it difficult to track particular documents or records.	Systems for EHS document control and record-keeping tend to be burdensome, bureaucratic, duplicative.	EHS document control and record-keeping systems are streamlined, simple to use and efficient.
Appropriate Scope/Coverage	The scope and coverage of existing approaches to EHS document control and record-keeping are difficult, if not impossible, to determine.	Existing systems for EHS document control and record-keeping tend to focus mainly on compliance-related information.	Existing EHS document control and record-keeping systems address the full range of relevant EHS-related concerns, including legal compliance.
Consistency/ Reliability	Availability, accessibility and quality of existing EHS documents and records is ad hoc and highly variable.	EHS-related documents and records are generally available, accessible and of acceptable quality.	Well-managed EHS document control and record-keeping systems consistently meet a high standard of performance.
Degree of Automation	Little or no systematic application of information technology to EHS record-keeping.	Multiple information technology systems are in use; applications are not coordinated or integrated.	Effective, streamlined use of information technology to efficiently manage EHS records.
			Overall Rating:

Managing Information

Management Process Assessment and Rating

	①	② ③	④ ⑤
Process Attribute	Needs Improvement	Meets Threshold Expectations	Significant Strength
Appropriate Scope/Coverage	EHS processes that could benefit most from automation have not yet been identified.	Management Information Systems (MIS) is used for some EHS information management applications. More comprehensive review of how MIS could improve EHS management has not been conducted.	MIS is used effectively to support information needs across a broad range of EHS management applications.
Clear Ownership, Roles and Responsibilities	No clear ownership or direction to EHS-related information management strategy, planning, investment or oversight.	Some limited coordination among EHS and MIS staffs but not managed as a process. Data and system ownership accountability still needs strengthening.	EHS-related information management roles and responsibilities are clearly defined and coordinated as part of a broader process for managing EHS-related MIS strategy, planning and execution.
Degree of Automation	Little or no systematic application of information technology to EHS management needs and purposes.	Considerable use of EHS-related MIS applications but not subject to any larger strategy or plan for their coordination or integration.	Effective streamlined use of information technology promotes better, more efficient EHS management throughout the organization.
Cost Efficiency	Mainly manual storage/processing of EHS-related information. Staff productivity relatively low for a large investment of their time.	Multiple overlapping systems do not align with key management processes or allow users to gain any efficiencies from possible system integration.	EHS-related MIS resources are invested and managed to provide integrated, cost-effective support to business users. Outsourcing is used as appropriate.
			Overall Rating:

Resources

Management Process Assessment and Rating

	①	② ③	④ ⑤
Process Attribute	Needs Improvement	Meets Threshold Expectations	Significant Strength
Number and Qualifications of EHS-Related Staff	No formal assessment of EHS-related staff needs has ever been conducted. EHS staff were largely "volunteered," lack previous EHS background and training.	EHS-related staffing levels generally appropriate to current service needs. EHS staff have adequate expertise and training.	EHS-related staffing levels reflect streamlined work processes, zero-based budgetary justification. Strong EHS professional skills.
EHS Staff Deployment	EHS staff function is not aligned with the rest of the organization's structure, including that of other key overhead functions.	Deployment of EHS staff resources is not inconsistent with the overall organization's basic structure.	EHS staff function is aligned fully with the nature of the organization's overall structure.
Technical and Financial Resources	Current levels of EHS-related technical and financial resources are not appropriate to the organization's existing EHS positioning, policy commitments.	Current levels of EHS-related technical and financial resources generally align with the organization's existing EHS goals and objectives.	Current EHS-related technical and funding levels are fully aligned with the organization's EHS goals and objectives; reflect strong EHS input in budgeting process.
EHS Improvement Motivation and Incentives	Lack of line management accountability for EHS performance. Little or no management attention to EHS reward/recognition and disciplinary programs.	Line management is generally accountable for EHS performance, but managers' appraisal process is not uniformly effective in enforcing it. Adequate EHS reward/recognition and disciplinary programs.	Clear expectation of line management accountability for EHS performance, supported by strong managerial performance appraisal process. Active, effective EHS reward/recognition and disciplinary programs.
			Overall Rating:

Organizations

Management Process Assessment and Rating

	① ②	③	④ ⑤
Process Attribute	Needs Improvement	Meets Threshold Expectations	Significant Strength
Top Management Commitment	Senior management awareness and understanding of key EHS issues is generally low; little sustained attention is paid to them.	Senior management is generally aware of key EHS issues affecting the business. They display some initiative in addressing them and require regular information on performance.	Key EHS issues are fully integrated into business management processes. Senior managers have personal EHS performance goals. EHS status and progress are regularly reported to the board.
Line Management Accountability	EHS management is generally left to staff specialists. Line operating managers have little incentive to become involved.	Line organization is actively involved in EHS management activities. Line managers have personal EHS goals, but accountability is not always reinforced.	Line organization drives EHS management, and individual managers behave as if EHS is their personal responsibility. Accountability is reinforced appropriately.
EHS Reporting Arrangements	Senior EHS professional reports more than two levels down from the CEO, lacks sufficient visibility and clout to be effective.	Senior EHS professional reports no more than two levels down from the CEO, has access as needed to corporate senior management.	Senior EHS professional is an integral part of the corporate management team, enjoying high visibility and credibility.
EHS Roles and Responsibilities	Corporate and line operating units' EHS roles and responsibilities are unclear and overlapping.	EHS roles and responsibilities are clearly defined and documented. Responsibility assignments to specific organizational units and individuals may still not be well understood.	EHS roles, responsibility assignments are clearly defined and well understood throughout the organization.
Continuous Improvement	Organization lacks an overall TQM process or equivalent commitment to ensuring continuous improvement generally.	Existing TQM process has been extended to include EHS activities. Specific EHS process improvement efforts are completed or under way.	TQM approach is integral to the way EHS activities are conducted. Strong customer focus and demonstrated record of achieving continuing EHS improvement.
			Overall Rating:

REFERENCES

1. Hardin, Garrett. "The Tragedy of the Commons," *Science*. 1968, Vol. 162, p. 1243.
2. Hardin, Garrett. *Living within Limits*. New York: Oxford University Press, 1993, p. 217.
3. Hardin, Garret. "The Tragedy of the Commons, *Science*. 1968.
4. Tibbs, Hardin. "Industrial Ecology: An Environ-mental Agenda for Industry," Arthur D. Little White Paper, 1991.
5. Hardin, Garrett. *Living within Limits*. New York: Oxford University Press, 1993, p. 217.
6. Gentile, Deanna. "Ink Outlook: Steady Growth and Evolving Technologies," *Modern Paint and Coatings*. 1996, Vol. 86 (7): pp. 40–42.
7. Willson, John and Ladd Greeno. "Doing More with Less: Improving Environmental Management Productivity," *Prism*. Third Quarter, 1994, pp. 95–107.
8. Senge, Peter, et al. *The Fifth Discipline Fieldbook*. New York: Currency Doubleday, 1994, pp. 235–293.
9. Porter, Michael. "What Is Strategy?," *Harvard Business Review*. November/December, 1996, pp. 61–78.

10. Browne, John. "Addressing Global Climate Change," Speech presented by John Browne, Group Chief Executive, BP International, Stanford University, May, 1997.

11. "Environmental Agencies and Intel Complete Nation's First Ever Cooperative Pollution Prevention in Air Permitting Pilot Plant," *Business Wire*. September 26, 1995.

12. Royal, Weld. "It's Not Easy Being Green," *General Sales & Marketing Management*. 1995, Vol. 147 (7): pp. 84–90.

13. Birchard, Bill. "Make Environmental Reports Relevant," *CFO*. 1996, Vol. 12 (6): pp. 79–80.

14. Shelton, Robert. "Cutting through the Green Wall," *Across the Board*. June, 1996, pp. 32–37.

15. Shelton, Robert and Jonathan Shopley. "Improved Products through Design-for-Environment Tools," *Prism*. First Quarter, 1996, pp. 41–49.

16. Senge, Peter. *The Fifth Discipline, The Art and Practice of the Learning Organization*. New York, NY: Currency Doubleday, 1990, pp. 57–135.

17. This case study is based on discussions with Kate Fish, Director of Sustainable Development at Monsanto Company. The author extends his appreciation for her insights and candor. Another useful source, particularly for quotations from Robert Shapiro, Monsanto's CEO, was Joan Magretta, "Growth through Sustainability," *Harvard Business Review*, January/February, 1997, p. 84.

Further Reading

Maira, Arun and Peter Scott-Morgan, *The Accelerating Organization: Embracing the Human Face of Change*. New York: McGraw-Hill, 1997.

Porter, Michael, "What Is Strategy?" *Harvard Business Review*, November/December, 1996, p. 61.

Porter, Michael and Claas van der Linde, "Green and Competitive: Ending the Stalemate," *Harvard Business Review*, September/October, 1995, p. 120.

Schwartz, Peter, *The Art of the Long View: Planning for the Future in an Uncertain World*. New York: Currency Doubleday, 1991.

Senge, Peter, *The Fifth Discipline: The Art and Practice of the Learning Organization*, New York: Currency Doubleday, 1990.

About the Author

Stephen Poltorzycki is a Vice President of Arthur D. Little and a Director in the firm's Environmental, Health, and Safety Management functional practice. The primary focus of his work at Arthur D. Little is to assist clients with strategy, management and technical needs related to environmental, health, and safety matters. Mr. Poltorzycki has worked for a variety of clients in the telecommunications, chemicals, petroleum, aerospace, motor vehicles and parts, scientific and photographic equipment, and financial services industries. For these clients he has developed strategies to deliver business value, and assessed, designed and implemented management systems to assure compliance, manage risk and obtain competitive advantage. Mr. Poltorzycki joined Arthur D. Little after working as a litigation attorney with Union Carbide Corporation and Kelley Drye & Warren, where he gained extensive experience in the litigation and management of safety and environmental issues. While at Union Carbide Corporation, he managed and coordinated the technical and legal resources necessary to assess the origin and ramifications of the Bhopal incident. He received his B.A. with High Honors from Wesleyan University in 1976 and his J.D. from Yeshiva University's Benjamin Cardozo School of Law in 1979, where he was Notes and Comments Editor of the Law Review.

Stephen Poltorzycki, Vice President, EHS Consulting, Arthur D. Little, Inc., Acorn Park, Cambridge, MA 02140-2390